PRAISE FOR S

'More than any treatise on black h. fiction, *Seven-Tenths: The Sea and its Thresholds* by James Hamilton-Paterson was a book from which I intended to sip, but ended up draining in one, long, spellbound draught. This is a series of maritime essays embracing topics from piracy to the invisible denizens of the depths. Personal, quirky, scientific, and poetical by turns—*The Songlines* of the sea.'—David Profumo, *Daily Telegraph*

'James Hamilton-Paterson has created an entirely original book about the sea, a subject which has clearly dominated his mind for many years. At first the reader is puzzled by the seeming shapelessness of this mélange, but is then carried along as if by a swift current because the writing is so felicitous, the navigation so impeccable, and the landfalls so interesting.'—Richard Hough, *The Times*

'*Seven-Tenths* is a meditation upon the sea and its meanings, whether as a dwelling place for sea gypsies among the Philippine archipelagos, a laboratory for suburban oceanographers, or an image bank for unconscious fears. As a diver, at home with the biology of coral and the physics of light perception, James Hamilton-Paterson knows that our fears are denied by the actuality, but as a novelist he logs the destruction of the last wilderness and its place in the imagination with the intense vision of an angry Thoreau.'—David Buckley, *Observer*

'The book provides an urgent reminder that the old sailor's hymn needs updating. It is not for those in peril on the sea we should be worrying. The sea itself is now in peril. *Seven-Tenths* should be required reading for the decision makers of every maritime nation.'—Alexander Frater, *New York Times Book Review*

'Under the sparkle and often superb imagery lie austere Conradian depths, most memorably in the description of the battleship *Arizona* rusting undisturbed among the flicker of damsel fish, a cenotaph to the dead of Pearl Harbor no sculptor could create. In a Conradian sense too the sea for the author is a mirror, a reflection of mankind's dreams and greed. The message is there among the glittering flotsam

and jetsam of the book, if humbug legislation continues to allow the dust-bowling of the oceans we may be writing our own epitaph as certainly in the long run as by the destruction of the rain forests.'
—Christopher Wordsworth, *Sunday Telegraph*

'*Seven-Tenths* is the kind of book that, were he alive now, Shelley might have written—the scientific curiosity, the lyrical interpretation of natural phenomena, the fascination with water and the sense throughout of being drawn down to the depths. Its technical terms and virtuosity might inhibit some, but it is a book full of references, footnotes, speculation and personal experience that stretch one's knowledge and haunt the imagination.'—Charles Sprawson, *London Magazine*

'British novelist and explorer James Hamilton-Paterson, who has been called the Bruce Chatwin of the high seas, seems to have set out to write a long meditation on the deep, its nature and moods, its effects on people. His book gradually turns into a quietly furious indictment of a fishing industry caught up "in a mad scramble to cut its own throat" and of all the other mindless human abuses of our oceans. Not since Rachel Carson have I heard such eloquence, such passion for the sea and everything it means to us. I was lucky enough to read the book while crewing on a sailboat off the coast of Maine, but could be in Kansas and still feel the magic.'—Michael Kernan, *New York Review of Books*

SEVEN-TENTHS

James Hamilton-Paterson

SEVEN-TENTHS

The Sea and Its Thresholds

Europa
editions

Europa Editions
116 East 16th Street
New York, N.Y. 10003
www.europaeditions.com
info@europaeditions.com

Copyright © 1992, 1998 by James Hamilton-Paterson
Preface © Copyright 2007 by James Hamilton-Paterson
First Publication 2009 by Europa Editions

Library of Congress Cataloging in Publication Data is available
ISBN 978-1-933372-69-3

Hamilton-Paterson, James
Seven-Tenths

Book design by Emanuele Ragnisco
www.mekkanografici.com
Cover photo © Robert Landau/Corbis

Prepress by Plan.ed - Rome

Printed in Canada

CONTENTS

For my Mother

Also
to the memory of
Ben Chong and Arnel Julao,
last seen on 20 December 1987
hoping to stow away aboard M/V Doña Paz
in order to join their families
for Christmas

Preface to New Edition

Re-reading this book was both reassuring and sobering. It was reassuring to an author that so little needed to be changed after fifteen years, and sobering for the same reason. In that time marine science has moved on and fascinating new data have emerged; but nothing, I think, to alter radically my original standpoint. On the political front, one signal and welcome development has been the ban by both the UN and the EU on drift-net fishing. Nevertheless, the steely attrition of the world's oceans proceeds at an ever-accelerating pace. Given the examples of well-monitored commercial stocks collapsing from over-fishing, the devastation currently being wrought by fishing fleets around the world is a spectacle that makes one wonder afresh at the hubris in styling ourselves *Homo sapiens. Homo stultus* would have been more accurate taxonomy.

In the long run, of course, our present cavalier treatment of the oceans matters not a jot. However heedless our self-interest, the sea will survive. It will outlive us and remain the source and origin of the planet's biota aeons from now. In any case, in writing this book I had no interest in producing an environmentalist jeremiad. On the contrary, I wanted to convey something of the affection, awe and often loathing that people have always felt for this mysterious body of water, both our cradle and the earth's largest geophysical feature. The sea leaks unconsciously into even the most land-lubberly of us from time to time, and manages to leave its tide marks in cultures as in rocks the world over.

The few minor changes I have made to the original text for this new edition are nearly all updatings, and those mainly to the chapter on fishing. I have also added some footnotes to take account of things like new free-diving records or oceanographical data. The single new addition is the inclusion of 'Sea Burial,' originally published in *Granta*, which describes a sea voyage I once made in the tropics. Neither precisely reportage nor fiction, it was an attempt to express that strange borderland where the sea's power to haunt sometimes seems as real a force as a wave's kinetic energy.

James Hamilton-Paterson
2006

Acknowledgements

I gratefully acknowledge the help given me in the writing of this book by the following:

Anna Badini, Anabel Briggs, Paul Brown, Peter Chaplin, Mario and Roberta D'Itria, Frank Donn, Earthtrust (Hawaii), the scientists and crew of the R/V *Farnella*, Stephan Feuchtwang, Bill Foster, Luis Go, Abdurrahim Hasim, Ellis Hillman, Frank Sionil José, Abdurahim Kenoh, Cdr Reuben S. Lista (Philippine Navy), George May, James May, John May, Anita McConnell, Gilbert Tait, Saladin S. Teo, Marcello Vanni, Ian West.

I would like to thank the Institute of Oceanographic Sciences, Deacon Laboratory, and especially Mike Somers and Tony Rice for their courtesy and help. Above all, Quentin Huggett has been unfailingly patient and generous with his time, a prodigality which it is a pleasure to acknowledge.

My agents deserve—and get—my very warmest thanks for their tireless support and help, as does Patricia Reynolds.

This book's commissioning editor was Richard Cohen, and I am conscious that it owes everything to his original encouragement. To that extent it remains his. The task of taking it on fell to Neil Belton, my present editor. I could not have wished for a more sympathetic and astute interpreter of a project whose intentions must initially have appeared idiosyncratic and opaque.

Finally, my gratitude to Mark Cousins is not easily expressed. One hardly thanks a friend for friendship, but one might well wish to put on record the affectionate recognition of a great debt. The entire book was written very far from his scrutiny, but scarcely a line without awareness of his presence. Private congruences apart, there was for me an especial punctuality about his public lectures at the Architectural Association in London, 1990–1, which had a profound effect on this book, just as the lectures had on all who heard them.

The above persons are exonerated from all responsibility for any errors of fact, judgement and taste, which are entirely my own.

Illustration acknowledgements

1. from *Zetetic Astronomy* (1873) 2nd edition.
   British Library, London
2. Private Collection
3. (i) & (ii) from *The Structure and Distribution of Coral Reefs* by Charles Darwin, (1889)
   (iii) & (iv) from *Corals and Coral Islands* by James Dana, London (1872)
4. from *The Christian Miscellany* (1860)
   Mary Evans Picture Library
5. from *Twenty Thousand Leagues under the Sea* by Jules Verne, (1870)
   Mary Evans Picture Library
6. *Miraculous Draught of Fishes* by Gustav Doré
   Mary Evans Picture Library
7. from *Concise Oxford Atlas* (1952)
   Oxford University Press

SEVEN-TENTHS

I am lost . . . These are the words the swimmer addresses in panic to the sunny universe into which he emerges, blowing water, disoriented. Ten minutes ago, perhaps twenty, he had set some fishing lines and slipped over the side of his tiny craft—a wooden insect with two bamboo outriggers—with a cord tying its prow to his ankle. He had been lying face down in the ocean, sun on his back, staring through the first thirty of a thousand metres of water. In the tropics these upper waters are flooded with light. Bright spicules drift past his eyes, crimson and electric blue, the jewelled phytoplankton streaming about the globe performing infinitesimal acts of chemistry which, much multiplied, succour all earthly life. By swimming down a couple of dozen feet he can look up and view other creatures from below: a shoal of garfish (whose bones are bright green) so high up their backs graze the rumpled mirror of air, the occasional flying fish breaking out and vanishing. The swimmer reflects on this mirror, imagining the sky weighing down on the sea and the sea holding up the atmosphere, curious about what exactly can be happening at the interface. If it were possible to magnify the activity, surely a buzzing skin of molecules would be revealed? Water molecules and air molecules so intermixed and saturated with atoms in common it would be undecidable which medium they constituted. At what point did these milling particles become waves? The swimmer loses himself in this quantum pun, in his speculations about boundaries, then suddenly an awareness breaks in that something is missing. There is a steady tug at his ankle, but too light. The long cord trails downwards, still firmly knotted to one foot. It is the boat which has gone.

His first act of panic is to spin in the water while trying to stand up in it: once, twice, three times, quartering a featureless horizon. Nothing. He is anchored by twenty feet of thin abaca hemp to the ocean. His masked face rams back through the surface as if by a miracle of misplacement he might discover the boat floating at ease in a fourth dimension some fathoms below. Nothing. The cord hangs down like the corals called sea whips, slightly kinked, whiskers of fibre standing out clear in this awesome lens right to the bobble of the knot in its end. The word this knot transmits through the water is 'adrift'.

The swimmer jerks his head up into the air again. Everything is plain. It is not possible, yet the boat has gone. I am lost. Panicked, he pants and spins, boatless, landless, and with the visceral ache of pure fear at what he has abruptly become: all alone and floating in the Pacific Ocean. Reason attempts to be reasonable. How far away could a boat possibly drift on a windless day? Also, eye level is barely six inches above sea level; a boat whose freeboard is little more than three times that could easily be hidden by the least swell. It is no doubt bobbing in and out of visibility even as he happens to scan the wrong horizon ... In any case, something altogether calmer is taking over: a lassitude, a fatalism whose roots reach back not to the beginning of his own life but, like the rope on his ankle, down into the sea itself. The twisted fibres, like ancient strands of DNA, connect him with vanished deeps, to primordial oceans lying in different beds. If he is lost now it is because he was already lost before ever setting foot in a boat, before infancy itself. He has no proper existence at all, being only a tiny hole in the water shaped like the lower two-thirds of a man. There is no way the tons of ocean can be held apart and prevented from filling the mould.

Yet it is not possible to give up, to go within minutes from being fit and happily occupied to renouncing life as if fatally injured. Fear returns in cycles. Looking around the liquid wastes beneath a brilliant sky he is set upon at intervals by the adrenal thought: This cannot be happening to me. . . . But it is. Then for a little while it is not; and in between assessing his chances of death by drowning, shark attack or exposure, into the swimmer's mind comes a sharp, vainglorious image of his predicament. Lacking all coordinates, he sees his own head occupying a fixed place. He pictures it sticking out of that expanse of curved blue ocean, a little round ball burnished with sun like the brass knob on top of a school globe. In his moment of loss he becomes the pivotal point about which the entire Earth turns.

1

Charts and Naming

The world according to zetetics: not a globe but a disc of geographical features bounded by a plane of infinite—or at least unknowable—extension and thickness. The hatched outer ring, all of which represents south, marks the insurmountable Great Ice Barrier.

I

T hey are a blank to most people, these bald oceans, much as they appear on a globe. As such they have their own coherence as two-dimensional representations of *not-land*. 'The sea' survives elsewhere in piecemeal images, scattered pictures which never link up. They include beach scenes from summer holidays, an aunt being seasick, storms, sunken treasure, pirates, monsters of the deep . . . 'The sea' as a reservoir of private imagery and public myth remains on the one hand without limits, on the other banally circumscribed. Yet it haunts us. We are full of its beauty, of that strange power it gives off which echoes through our racial history and fills our language with its metaphors. The salt which is in seawater is in our blood and tears and sweat. The lungs of an infant *in utero* can be seen rhythmically breathing as it inhales and expels amniotic fluid, even as its oxygen supply comes from the mother's bloodstream via the umbilicus. Each of us has breathed warm saline for days on end and survived. The lungs themselves derive from fused pharyngeal pouches, and branchial clefts ('gill slits') still form temporarily in all chordate embryos, including humans, reminding us that something which became *Homo* did crawl up a beach many millions of years ago. The satisfaction for certain people of walking back down a beach and into the sea is akin to that of a long-postponed homecoming.

Too late, though. We have lost our place and no longer know how to return. It is never quite enough, the ecstasy of splashing or the torpid floating while a fine slick of tanning oil spreads its iridescent pollution around us. Too much knowledge, too much strangeness slips by. Maybe we should approach the whole thing through science. Marine biology will tell us about the phytoplankton in their gemlike flakes, trigonometry will reveal where we are. The business of orientation on the blank and shifting waters of the open sea, of establishing a fixed *point de repère* which is not a landmark, is central not merely to navigation but to various sciences coming under the general heading of oceanography. (The very word suggests a difficulty, the writing down of an ocean.) Until a certain moment in history there must have been a conceptual impossibility in the idea of a sea chart without a coastline on it, implying as it logically would the drawing of lines and boundaries—albeit notional ones—on a fluid surface. Being boatless and lost in mid-ocean at noon in the tropics at least makes vivid certain problems which have faced all navigators and cartographers, with the sun directly overhead and the seabed far out of sight a mile below. The panic of a careless swimmer keen to avoid joining his ancestors thus makes a good starting point as he twirls despairingly round and round like a demented compass needle in search of any bearing, any point of reference, any direction other than down.

The swimmer wanted to know many things. Gazing straight down beyond his parsnip-white feet into the purple depths he wanted to know not just what was down there but how it might be mapped. How did one make an accurate chart of the seven-tenths of the Earth's crust hidden from sight? Since he had always wanted to know such things he wondered why he had never become a marine scientist. This was imponderable so he inverted the question and speculated about what it is that makes a scientist choose his particular field of study. This swimmer, who is almost as obsessed with the idea of the sea as with its actuality, wondered if oceanographers necessarily shared his own obsession, and if so, to what effect. Did they love to go delving into the sea? Was it scientifically useful to be imaginatively caught up in the deeps?

Such speculations bring me in due time (early December 1990) to Pier 40 of Honolulu Port and aboard the R/V *Farnella*, a British-registered research vessel. The *Farnella* is under charter to the United States Geological Survey, which is using her to map the seabed contained by the US 200-mile Exclusive Economic Zone. It seems curious that this grand national task should be carried out using a system developed in Surrey and mounted aboard a converted Hull trawler. Some years ago the ship's refrigerators were taken out and her forward hold filled with huge spools of cable for towing instruments. The stern hold is now largely given over to a spacious air-conditioned laboratory humming with bays of computers, sonographs, plotters and assorted electronic gear. When we leave Honolulu we have aboard fourteen crew, thirteen scientists (American and British) and myself, my capacity officially listed as 'Fly on wall'.

We spend the first two days at sea simply reaching the area of the survey, which is that surrounding Johnston Island. At this moment Johnston Island is a high-risk, high-security area since this is the atoll where the US is destroying its stocks of nerve-gas shells and other chemical weapons from bases in Germany. Several of the scientists had put ashore there briefly on a previous trip. 'Can't land if you've got a beard,' said one, 'or even a couple of days' growth. Your gas mask mightn't be airtight.' 'Don't worry,' said another. 'We'll be over a hundred miles away most of the time and with any luck upwind.'

I wander the ship, stare at the ocean rolling past, meet the Captain. 'He's temporary,' a crew member tells me. 'Wicked sense of humour.' A dapper man, he had distinguished himself during the last Cod War of 1975 by deliberately ramming an

Icelandic trawler. Otherwise, his sense of humour seems not greatly in evidence. Ah, but a previous captain, now, he'd been gross in all sorts of ways. 'Pig fat, he was. Also, I remember up on the bridge, you'd never know when, sort of absentminded-like he'd take out his teeth to scratch the eczema on his legs, the top plate usually, and you'd see this dust like cornflour come puffing out of his trousers. Then he'd slippem back inna-gain.' The sea has always been kind to eccentrics, and indeed the *Farnella*'s crew give off a feeling of dourly amiable toler-ance, even towards scientists and flies on walls.

The main piece of equipment for bathymetry aboard, and the chief tool of the entire USGS programme, is GLORIA: Geological Long Range Inclined Asdic ('asdic' being the British equivalent of the American 'sonar'). Since seawater is four and a half times better than air at conducting sound, some kind of sonar system remains the most effective way of map-ping a seabed which may be anything up to 11 kilometres below the surface. The very earliest sonars were simple echo rangers. The sinking of the *Titanic* in 1912 had made finding some sort of defence against icebergs an urgent matter, but it was actually World War I—and specifically the development of submarine warfare—which encouraged sonar technology and in so doing gave the biggest boost to modern oceanogra-phy.[1] One off-shoot of military research was an efficient echo sounder for cable laying.

In 1854, Lieutenant Matthew Fontaine Maury of the US Depot of Charts and Instruments had published a profile of the Atlantic seabed, his *Bathymetrical Map of the North Atlantic Basin with Contour Lines Drawn in at 1,000, 2,000, 3,000 and 4,000 Fathoms*. This was based on only a couple of hundred deep soundings achieved with a weighted line, and its imposing title suggested a survey rather more thorough than it actually was. Nevertheless, it interested a good many people,

---

[1] Asdic: Allied Submarine Detection Investigation Committee.

among them a rich industrialist, Cyrus W. Field. Field had long dreamed of linking America with Europe by telegraphy. The seabed between Newfoundland and Ireland was, according to Maury's survey, mostly plateau. Suddenly, the romantic notion of voices emerging from the deep took on real commercial possibility. In 1856 the Atlantic Telegraph Company was founded and two years later it laid the first transatlantic cable. Unfortunately, Maury's 'plateau' turned out to include over 1,000 miles of fracture zone: the spine, in fact, of the Earth's largest mountain range, the mid-Atlantic Ridge. This runs for 7,000 miles with peaks rising two and a half miles above the seabed. Within three months the cable had broken. Under the continued pressure of the Company's commercial determination, though, Maury's team were inspired to develop better sounding devices and bathygraphy. The result was a permanent telegraph link, finally laid in 1866. Yet even the best sounding techniques using plummets and lines had great limitations (see Chapter 6). At least another half-century was to elapse before the novel alternative of sonar was devised.

In 1922, a historic moment in cartography and oceanography occurred when the USS *Stewart* used a Navy Sonic Depth Finder to draw a continuous profile across the bed of the Atlantic, sixty-eight years after Maury had published his map. But whereas Maury's had involved large amounts of guesswork, the *Stewart*'s involved none, aside from certain problems of interpretation. Otherwise the soundings were fast, accurate and simple to take. The principle of sonar is straightforward. A pulse of sound is bounced off the sea floor and received by the vessel which sent it. Having made small allowances for such things as the speed of the vessel, it is a matter of the simplest arithmetic: divide the time elapsed by two and multiply the result by 4,800, which gives the depth in feet, 4,800 feet per second being the mean speed of sound in water. (The layman is astonished only that, with a principle as thoroughly understood and long turned to practical use, it should have taken so long to develop radar, that apparently magical World War II conceptual breakthrough. Since then, with a good deal of technical wizardry but zero intellectual daring, radio pulses and even laser beams have been bounced off the Moon, ranging it to a matter of a few centimetres.)

There are, however, certain difficulties to this kind of bathymetric depth sounding. One can be intuited from the phrase 'mean speed of sound in water,' since in practice the speed of sound in seawater varies. It travels faster, for instance, if the temperature, pressure or salinity increase. This is to leave aside for the moment all the difficulties in interpreting the returning signals, which may be severely affected by shoals of fish and other powerful sources of scattering. A further drawback is that a series of pulses focused vertically downwards beneath a ship's keel will give only a series of depths. In order to build up a contour map of how the sea floor actually looks a ship using this method would have to make a laborious succession of very closely spaced passes. Otherwise, only a narrow cross-section of the seabed being traversed is possible.

A way around this difficulty was pioneered at the Institute of Oceanographic Sciences in Surrey. Instead of using vertical echo sounding, they developed a sidescan sonar in which the sound pulses are emitted sideways in two slanting fans, one on either side of the ship and at right angles to its course. The outer edges of these fans brush irregularities on the seabed as far away as 30 kilometres. The returning echoes, since they do not come merely from a single point directly beneath the vessel's keel, are complex and take skill and experience to interpret. Yet the profile which emerges from the plotters provides by far the clearest imagery yet possible of a continuous swathe of seabed, certainly on a cost-effective scale. GLORIA's efficiency is striking. In good conditions and with the ship travelling at 8 or more knots it can reveal a strip of seabed 60 kilometres wide. In twenty-four hours this amounts to mapping more than 20,000 km², or an area the size of Wales.

This cruise in the *Farnella* is one of a series at the end of a great project on the part of the US to map the 200-mile EEZ around all 19,924 km of its coastline. The EEZ of the continental US has already been mapped, including Alaska and the Aleutian Islands. Now *Farnella* is working away at the Hawaiian Chain. As the scientists aboard keep telling me, only the GLORIA system could have covered such an area, and even that has taken ten years.

Such things are explained as the ship bounces and judders through seasonal Pacific rollers on her way to the survey area. On and off a tropic sun blazes the sea to a luminous indigo across which flying fish skip and glide. In the intermittent bursts of sunlight there is a stampede of scientists up on to the deck where they grill themselves on towels, redolent of coconut oil. December! Hawaii! Field trips sometimes throw in for free the costly ingredients of other people's holidays. They are watched by two boobies clumsily slithering on the button atop the radar mast while a small albatross flies elegant, mournful rings around *Farnella*. On successive days of watching I never once saw it move its wings, only soar and tilt in any direction and at any speed into the headlong breeze while keeping pace with the ship. It looked as though it knew by heart a map of the ocean's surface, a map no man will ever make.

Time was spent checking the various instruments which were soon to be lowered into the sea and towed behind us. First of all GLORIA itself: a large yellow torpedo lined inside with banks of transponders precisely angled to give the correct fan-shaped pulse. It sits in a hydraulic cradle directly over the ship's stern. From time to time technicians climb up to tighten a nut and pat it protectively. Even without its cable it is worth nearly half a million pounds. There is spare cable but only one GLORIA aboard. 'We don't even like to think about that,' is the response to the obvious 'What if?'

Amidships, two little 'fish' like fat yellow bombs wait on wooden trestles. These are the 10 kilohertz and 3.5 kilohertz sonars. The first will act more or less like an old-fashioned depth sounder to give the distance to the seabed immediately beneath us. The second will send pulses up to 400 metres into the ocean floor to provide some idea of the underlying geology. It is explained that its readout, taken by itself, would be fairly meaningless since in physics, as elsewhere, one cannot get something for nothing and the deeper penetration gained by using a higher frequency is at the expense of detail and spread. But taken in conjunction with information coming from other instruments this data will be usefully corroborative.

Over the starboard stern will go the air gun. This is a heavy pot of machined steel, something like the body of a pneumatic drill and similar in principle. Connected to a high-pressure air hose it will 'fire' itself every ten seconds with a loud detonation and send sound waves capable of penetrating up to 1 kilometre into the ocean floor. This is the oceanographer's equivalent of seismologists' 'thumper trucks' which used to drive around the deserts of the Middle East, dropping huge blocks of concrete and listening for signs of oil-bearing strata. The air gun is a safer and simpler alternative to throwing overboard dynamite charges at timed intervals. Its echoes are received on a hydrophone 'streamer,' hundreds of metres of transparent plastic tube filled with bundles of multicoloured sensor wires and light oil. The streamer may be towed well over a kilometre behind the ship and is of all the equipment the most susceptible to damage. (At the end of the cruise when the streamer is wound aboard there is a section leaking oil with a broken shark's tooth embedded in the gash.)

Over the port stern will go a magnetometer to measure magnetic variability in the Earth's crust, and down in the lab is a gravimeter to record differences in its gravitational field. This machine looks, and is, expensive. It is suspended in a cradle mounted in computer-controlled gimbals, dipping and tilting so it appears to be the one thing in the lab which is constantly in motion, whereas it is really the only thing aboard remaining utterly still while the ship gyrates about it. At supper the conversation turns to where might be the best place on Earth for setting high-jump records, a particular spot with significantly weaker gravity. All the best ones seem to be covered by a couple of miles of water. In response to a remark of mine which betrays real ignorance about gravity, Roger says kindly:

'I suppose one always imagines the surface of the oceans as basically flat. Ignoring waves and local storms, of course— they're just "noise". But apart from its being curved to fit the surface of the globe, one thinks of the sea as having to be flat because at school we're told water always finds its own level so as to be perfectly horizontal. On a small scale that's pretty much true, though when I was about ten I remember being surprised when someone pointed out that all rivers are tilted, and if you row upstream you're also rowing uphill as well as against the current. Anyway, since gravity varies from place to place it acts variably on the sea, too. When you start using instruments like the ones aboard this ship you really appreciate how the ocean surface actually dips and bulges all over the place. It shows up best from space.'

He explains that by having enough satellites in orbit making passes over the same area, day after day for months on end, it was possible to build up a mean reading for the height of the sea's surface at that spot. It took a long time because there was a good deal of 'noise' to be discounted: wind heaping, sudden areas of low atmospheric pressure which could suck the sea upwards as if beneath a diaphragm, even very low-frequency waves with swells so long they might take half a day to pass. But if the satellites went on measuring the same spot for long enough such fluctuations would even out and a geodetic point be established: a mean distance to the sea's surface as measured from the centre of the Earth. By building up enough geodetic points it soon became clear that the oceans were anything but flat.

'What's more, if you match this up with the underlying features on the seabed, you'll find that the surface of the sea broadly mimics the topography underneath. And the reason for *that* is fluctuations in gravity, which depends on the density of the crustal material.'

It is a pretty notion, that the sea follows the Earth's crust like a quilt laid over a lumpy mattress. It is also odd to think that to some extent the depths of the oceans can be read from space. Over a plate of steak and kidney pie (the galley makes no concessions to the tropics) I presume this means that sometimes a ship has to go uphill and downhill.

'Certainly. But the "hills" are so slight you'd never know. We're talking a few tens of centimetres here, spread out over kilometres. Sure, a ship often has to go uphill, though it won't be using any more energy. All points on the hillock's surface have the same gravitational potential, obviously.'

This is not obvious to me, and nor is it any more so after Roger has explained it several times in different ways. I tell myself that physics is humiliating not when it defeats the intellect but when it confounds the imagination. This makes me feel better. Giving up on me, he reverts to a sort of 'Ripley's Believe It or Not' mode suitable for lay company. Roger is himself a geologist and in describing the planet gives the impression of talking about a beach ball underinflated with water: labile, plastic, sagging and crinkling and bulging. It is not only the oceans which respond tidally to the Sun and the Moon; the Earth's surface does as well, rising and falling twice a day.

When the Moon is directly overhead it is pulled up by half a metre.[1] What is more, this elastic crust seems to have a frequency of its own at which it resonates. A Russian geologist, S. L. Soloviev of the Moscow Institute of Oceanology, recently made seismograms of micro-earthquakes under the Tyrrhenian Sea. Using bottom seismographs (developed from nuclear explosion detectors originally designed to enforce the Test Ban Treaty), Soloviev began picking up a distinct, ultra-low frequency oscillation which he thought was most likely the fundamental frequency of the Earth's crust itself.

That night I go to bed with my head full of marvels. In the course of the evening I had also learned that the sea levels at either end of the Panama Canal were different by nearly half a metre, and the same went for the sea on either side of the Florida Peninsula. This was caused by things such as the heaping effect of wind and the Coriolis force. But I am most captivated by the idea of the Earth's crust vibrating at an ascertainable frequency since it would theoretically be possible to calculate the precise note. True, it would probably not be a pure tone because there would be all sorts of harmonic interference from irregularities such as mountain ranges. Yet it ought to be possible to determine the fundamental note of the planet, the music of our spheroid.

I also wonder at the notion of the sea's surface modelling the plains and mountains, chasms and basins beneath the keel. It is not hard to believe this at the moment since we are all being thrown about our bunks by the *Farnella*'s plungings as if she were ploughing across rough country. We have reached the

[1] The extremist pull is, of course, when the Sun and Moon are perfectly aligned, as in the total solar eclipse of July 1991. Astronomers observing this phenomenon from the top of Mauna Kea, an extinct volcano in Hawaii, noticed a minor eruption in nearby Mauna Loa, believed until then to be equally extinct. This is considered highly suggestive of there being 'tides' in the Earth's crust as well as in its seas.

foothills of the Necker Ridge. More than two miles below us a mountain chain thrusts steeply upwards. I bang about in my wooden trough and tell myself this is just 'noise'.

Next morning the sea is quieter. The ship heaves to and a certain tension comes over the scientists as one by one the precious instruments are carefully deployed. Launchings and retrievals are the moments when damage is most likely and although there are several workshops aboard for mechanical and electronic repairs our sailing patterns are planned to the nearest nautical mile for the next several days. Worries revolve around personal responsibility for the correct functioning of machines. The spectre of disgrace and delay flits about the ship until everything is safely in the water and the test readings are monitored. In the end these anxieties are probably rooted less in codes of professionalism than in the huge expense of modern oceanography. Great sums of money are being lowered delicately into the ocean. Any delays would be tantamount to damage, chunks of money becoming dislodged and drifting down to the seabed where they would dissolve and be lost for ever. Even the crew seems less laconic while all this is going on. When we are under way again there is a feeling that the *Farnella* is more in the hands of scientists than seamen and the crew can now be found in odd corners reading copies of the Hull *Daily Mail* which were flown out in bundles to the shipping agent in Honolulu. Shipboard life settles into routine. It is still curious to be in the Pacific in a British ex-trawler with a television in the lounge showing video-cassettes of highlights from last season's Hull Kingston Rovers matches. Not to mention the cuisine. At the same time we are, as an IOS zealot proudly says, 'at the leading edge of geophysical seabed surveying'.

The issue of who is really in command of the ship is interesting, as is the whole idea of a joint survey paid for by the US

Government using a significant proportion of British equipment and scientists. In a legal sense the Captain has full and final responsibility for the ship. Yet it soon becomes plain that his actions are largely determined by the exigencies of the survey, which is costing the American taxpayer such a pretty penny. It is the USGS which has chartered the vessel and hired GLORIA and so calls the tune. On the other hand the scientist formally in charge of this particular cruise is a Briton, one of GLORIA's original developers at IOS. At the same time one of the young American women aboard is responsible to her government for completing this leg of the survey . . . All this interweaving of authority is glossed for me as 'A joint effort. Absolute cooperation and consultation. Democracy in action like you wouldn't believe.' This is emphatically not science for the sake of science, a matter of drifting about the Pacific like the old *Challenger* in the 1870s, sounding here and dredging there at whim. This is time-and-motion science, with a given area of blank map to be filled in a given time. And the whole issue, for very cogent reasons of physics, hinges around the matter of navigation.

As is all too clear to anybody swimming in circles looking for a lost boat in the middle of the ocean, one has no position in water. When mapping the seabed from a moving ship, therefore, accurate navigation is of crucial importance. Without the ship's position being known from one second to the next the most beautiful chart of peaks, ravines and plateaus would be useless. The only thing known would be that they were down there somewhere. Establishing the ship's course along lines as straight as possible (always allowing for the Earth's curvature) requires much work, not least because the swathes GLORIA maps must lie next to each other without gaps or wasteful overlapping. On the chart table in the lab is the dot of Johnston Island, a pencil circle whose diameter represents 400 nautical miles inscribed about it. High up in its top left-hand quarter a chord shows the first leg we have just started. Next to it is written the estimated time at which we should come about for the return pass, each leg getting longer as we eat downwards into the circle. If all goes well, by the end of a fortnight we should have hatched off about a quarter of the total area.

While the lab computers flicker with the instruments' returning signals, various repeater gauges give the ship's speed through the water, its speed over the ground, the wind speed and any consequent degree of yaw. If to remain on a straight course against a quartering wind and current the *Farnella* needs to sail crabwise, GLORIA's angle will also be fractionally oblique to its correct path. The result is that its signals will no longer be exactly at right angles to this course and the map will be distorted. Information on all these factors is fed into the computers, which correct for them. In order to determine the ship's position at any moment the *Farnella* uses GPS or Global Positioning System. This depends on satellites and eventually, provided there are still spare slots in an already overcrowded geostationary orbit, the system will cover the Earth and in theory allow a person anywhere on the planet's surface to determine his position to within a few metres. This would not be of the slightest use to a lost swimmer looking for his boat.

Bored with the sight of bright red digital figures flickering their decimal points on display panels, I wander off in search of sound. Down in the forward hold, above the banging of the ship's forefoot into wave troughs, the chaffinch-like chinking of the ten kilohertz 'fish' can be heard through the steel hull. Up on the stern deck there is a sharp cracking sound every ten seconds, the higher frequencies of the air gun's detonations being transmitted back up the compressed air hose. In the water astern white puddles dimple and churn to mark the boilings of released air. They follow the ship with the measured pace of footsteps.

Very occasionally from a chance position down in the hull,

at some freak acoustical window, it is possible to hear GLORIA's peculiar yodel. The instrument emits a correlation signal; instead of a single bleep its pulses take the shape of a whistle which swoops up and down. This is so the echo will be unmistakable, the electronic ears listening for its return being tuned to exclude all other signals. Even so, knowing how to read the GLORIA trace as it emerges from the plotter is a matter of much experience. Since parts of the signal are making a round journey of sixty kilometres or so, while others may travel only five (that edge of the fan nearest the ship) the returning echoes become mixed up with the fresh outgoing pulses, even with still fainter returns from previous signals. There may also be leakage and scattering, with stray echoes reflected back down from the water's surface.

How different the *Farnella* is from the old *Challenger!* The real distinction between this kind of oceanography and all that went before is not merely that the technology has changed, and with it the techniques for analysing data. It is that the scientists themselves are using different senses. Nobody is actually listening to these signals returning from unexplored regions laden with information. The lab is filled, not with the hollow pinging familiar from submarine war film soundtracks, but with the click and whir of plotters and jocular bouts of repartee. No one now wears headphones and a rapt, faraway look, attentive in ambient hush. For all that modern oceanography relies so much on acoustic techniques, it is machines which do the listening. When I flip a switch on a panel which feeds through a tiny speaker the actual noise of the signals, the American technician Bob sets his face into that expression which in TV shorthand stands for displeasure. 'That sound gives me a headache,' he says. 'It's so goddamn monotonous.' I refrain from babbling about hidden subtleties, since they are still there; it is just that they are on an inked printout.

Allowing electronic devices to replace our senses while

reducing so much information to visual imagery must have its consequences. Generally speaking, under-used faculties tend to atrophy. It has long since become a cliché in the pages of *The Lancet* and the *BMJ* to wonder whether the old-fashioned, pre-war family GP with his training in how to watch, to listen, to smell, touch and even taste may have understood more about his patients' health than does his modern counterpart with his reliance on laboratory techniques and diagnostic machinery. Perhaps in dealing with the natural world at an electronic remove scientists in certain disciplines may also risk missing as much as they learn. How many naturalists nowadays have the artist's eye, like the great nineteenth-century scientists who so lovingly sketched their specimens in the field? It is not only sensibility but memory itself which atrophies, since the need for attentive observation is less. The camera takes the place of the eye, the recorder of the ear, the computer of the memory. A laconic finger on a keyboard summons up data, an image. Less need, less time now for Edward Lear's scrupulous parrots or Audubon's American birds, for the hundreds of sketches made aboard the *Challenger* or for anatomical drawings as fine as Jan van Rymsdyk's of the human uterus. Nor is there much call for writing that describes specimens as Philip Gosse described *Cleodora*, a tiny snail known as the sea butterfly, which floats in tropical oceans. 'A creature of extreme delicacy and beauty . . . The hinder part is globular and pellucid, and in the dark vividly luminous, presenting a singularly striking appearance as it shines through its perfectly transparent lantern.'[1]

[1]Philip Henry Gosse, *A Textbook of Zoology for Schools* (London, 1851), p. 220. Gosse was later criticised by his son Edmund in *Father and Son* for having seen 'everything in a lens, nothing in the immensity of nature.' Yet his descriptions were both lyrical and accurate. That his sense of wonder served to reify an avowed religious purpose (the Textbook was published by SPCK) is unimportant. What counts is that he wrote about each organism with an affectionate eye, as though seeing it for the first time: the precise quality which a textbook serves to annul.

Roger agrees that until well after World War II sonar operators and scientists had spent much of their time with headphones on, listening to the seabed and to anything else whose noises fell within humanly audible wavelengths. They gave their own nicknames to certain familiar sounds, especially those which resisted all identification. One of these became known as the 'North Atlantic Boing'.

'Now and again something uncanny happens which makes you wonder about what's down there. We've sometimes picked up signals which aren't GLORIA's own echoes but *imitations*. The electronic analysis is quite clear. They're definitely some kind of deliberate response. But what could possibly imitate a sound as complicated as a correlation signal? There are really only two possibilities: either a submarine or whatever creature produced the "North Atlantic Boing" while flirting with sonar operators. Sometimes even our other sonars—the "fish," for example—provoke an almost angry response. We've had the single bleep of the ten kilohertz being answered by a double bleep, or a triple, even a quadruple, as if something's deliberately mocking it.

'You've got to remember that sound travels well in water and we're making a godawful noise down there. We're sending out four different signals powerful enough to bounce off the seabed thirty kilometres away, two of them capable of penetrating hundreds of metres into it. That's a huge amount of sonic energy and we must be absolutely deafening, maybe even lethal, to some animal species. Certainly a whale would find us audible for hundreds of miles, perhaps a thousand.

'So who or what is the author of the "North Atlantic Boing"? What creature are we enraging or provoking to mimic us? I presume cetaceans. We haven't the first clue what goes on in a whale's mind. We don't even know what's in a *parrot's* mind.'[1]

---

[1] In 2005 the US National Marine Fisheries Service finally traced an identical Hawaiian 'Boing' to male minke whales marking their territory and attracting mates.

At this point a very sober, older scientist breaks gently in to refute Roger's account of the 'Boing.' He has the air of someone who has long ago signed the Official Secrets Act and has never regretted it. It makes his explanation oddly dark, vague with the pregnant ellipsis of someone in the know who cannot tell all he knows. It comes as a reminder that the military has its own interests in oceanography and a good deal of knowledge and technology overlap. There was never any such thing, he says, as the 'North Atlantic Boing.' The fact is, it was only ever audible on one particular occasion off one particular part of the east coast of the United States. Significantly close, in fact, to one of the major naval bases. What is more, GLORIA uses what used to be a naval signal on the same carrier frequency. Quite possibly it was a submarine communicating with an underwater beacon.

As for Roger's cetacean theory, there are a good few holes in that, too. All cetaceans are far-ranging and one would expect this 'Boing' to have cropped up in widely separated areas. Second, GLORIA's 6.5 kilohertz is in the 5 to 10 kilometre range, and why would any cetacean be interested in anything at that distance? As for the allegedly angry responses to the 'fish,' Roger will—he is certain—recall a good deal of active porpoise noise in the Bering Sea, especially at ten kilohertz. There was indeed enough energy to break through the correlator return, but it was continuous 'active' noise, not a genuine return signal, and it was also associated with biological activity in the water column. It could well have been caused by numbers of animals all fleeing the GLORIA sound at the same moment. Nor should we forget that porpoises can produce all sorts of noise and interference merely by breaking the surface of the water as they swim. Air bubbles are one of the richest sources of noise—look at the swim bladders of hatchet fish—and sudden frothing and foaming plays havoc with signals and can produce the oddest effects . . .

This is not dissension, of course. The man merely drifts amiably away, leaving behind a feeling of official cold water having been poured, as well as a suspicion that perhaps not everything has been neatly explained after all.

The next day it is discovered that existing charts of this area, which like most maps (other than those of pirate treasure) look completely authoritative, are quite wrong. Whole features are either absent or misplaced. We examine the printouts. Entire mountains flicker in and out of existence somewhere down there in the cold darkness. The bathymetry is all haywire. GLORIA knows best. There remains the experience of being present when a portion of the Earth's surface is discovered. This is a rare sensation for the layman in the late twentieth century, and the diminishing opportunities for experiencing it must belong almost exclusively to potholers and cavers, apart from oceanographers themselves. At this moment, except for the handful of people present in the *Farnella*'s laboratory, I know more of the range of hills we are traversing than anyone else in the world, be they professors, explorers, Nobel laureates, fellows of distinguished societies or captains of industry. In fact, I know more of this piece of America than the President himself, and I am not even a US citizen.

This schoolboy superiority is too brief and dubious to be at all satisfying. Moreover, it is imbued with a certain sadness. One more thing has fallen under *Homo*'s rapacious gaze, and as always the knowledge is not neutral. By its very nature this project makes one constantly aware of the question of ownership. Effectively, anything we find down there belongs to the United States which, by annexing its tranche of the seabed, is actually adding a prodigious 2.9 billion acres to the 2.3 billion acres of dry land it already commands: rather more than the same again. Quite apart from any military consequences, the economic aspects are plain. This makes the fixing of all EEZ boundaries a matter of great importance, both scientifically and bureaucratically. As soon as one starts to consider it in any detail, obvious difficulties present themselves. On this cruise we are supposed to be mapping the EEZ around Johnston Atoll. Are the two hundred nautical miles to be measured from the island's coastline or from its centre? There again, if they are measured from the coastline does the boundary have to describe a neat circle or should it faithfully follow every little promontory and indentation so that on a chart Johnston Island will eventually appear as a small blob in the middle of a dotted outline of its own hugely magnified ghost?

According to Roger everything depends on the size of the island. International law stipulates how big an island has to be before the EEZ may follow its coastline. There again, a mass of guidelines define a coastline and how to treat it. If the mouth of a bay is less than a certain width it must be considered as if a straight line ran across it. Such niceties are far from being purely academic. According to Colonel Gaddafi the Gulf of Sirte is entirely within Libya's territorial waters whereas the US Navy, which is constantly on patrol there, claims the Gulf is international.

'It usen't to be like this,' Roger says. 'When I first started in oceanography—and this is only about ten or fifteen years ago—you could still go where you liked to do your scientific cruises, just as they did in the last century. Nobody ever objected if a British survey vessel chose to do some sounding or coring forty miles offshore. Nowadays if we want to do fieldwork inside somebody's EEZ we have to apply to the Foreign Office, can you believe, to get permission on our behalf and like as not we'll have to agree to take appropriate foreign observers on board. It's a ridiculous hoo-ha.

'It's all about wealth, of course. Round here there are some huge fields of nodules that are potentially rich pickings. It's up to us to locate them precisely. But here, for example,' he points to a chart taped to the bulkhead, 'south-east of where we are now, that's the Clarion-Clipperton Fracture Zone, famously rich in nodules. As you can see, it's in the middle of the Pacific. It doesn't fall into anybody's EEZ. Yet in the early eighties President Reagan declared it to be an area in which the US would regulate all mineral mining licences. Incredible, when you stop to think about it. It's not even US territory. Reagan's argument was that there was a sort of power vacuum there and that *someone* had to supervise things otherwise there'd be all kinds of skulduggery. Pure altruism, you aren't thinking. So the whole thing has now been pretty much carved up by about four consortiums. No skulduggery there, naturally. Now, the question we're asking is, supposing somebody starts mining manganese nodules in an area like that and you come along wanting to do your bit of oceanography? Neither they nor you have an obvious legal claim. It's international waters. But I somehow can't see a deep sea mining outfit sitting idly by while a fully equipped research vessel heaves-to a mile away to do some close investigating of the field.'

One afternoon I spend time in the dark room where Roger is making prints of the laser printer films of GLORIA scans. It is drudgery but needs constant vigilance. The contrast between the high technology outside in the lab and the low in the dark room is marked indeed. The ancient printer is suspended on a wooden tray by four cords from the ceiling. The scale of the prints changes fractionally all the time and focusing is critical. The reason is the fundamental cartographer's problem of how to represent a curved surface on a flat one. On the variant of the Mercator's projection the Survey is using the latitude lines grow further apart the further north one goes. At the levels of accuracy required, the *Farnella*'s slightest northward drift can make a difference.

'It's not for fun, all this,' reminds Roger's voice in the darkness. 'It's about potential megabucks. We're likely to find nodules hereabouts. If so, somebody else will come along on a sampling trip to see what the quality's like and they'll want to be able to arrive at the precise point and drop a sonar buoy. And they'll be doing it from our GLORIA scans. I know it's a bore, fiddling about with the focus and scale and stuff, but I do quite like that it's a tangible effect of ordinary physics . . . Hey, you know the Forth Bridge? Well, its two towers are absolutely plumb vertical, but their tops are a centimetre further apart than their bases. Curvature of the Earth. I love that, don't you?'

The enthusiasm of his voice in this stuffy cubicle full of the reek of developer and fixer is contagious and banishes the monotony of the task. He is apologetic in case his recently completed doctoral thesis sounds dull. 'Actually, it's on acoustics and sedimentation . . . Basically, why is it that GLORIA's signal comes back at all, rather than simply scattering away all over the seabed?' While still young, Roger is a veteran of field trips all over the world, from Tahiti to Alaska. One of the surprises of modern oceanography has been the discovery of undersea fans and river systems which, as mapped by GLORIA, bear a great resemblance to aerial and satellite images of river deltas and similar land features. This, too, has a commercial aspect. 'Take gold. A glacier shield like the one north of Alaska liberates gold from the Earth's crust as it grinds away. This, in the sediment, eventually flows into an undersea river system. Over millions of years that's a lot of gold. But at the same time the crustal movement shifts these ancient deposits about while the actual sources don't necessarily change position. So an old auriferous sediment may today lie deep beneath a shallow modern fan system containing practically no gold at all.

'Now, if you've been looking at GLORIA and seismic scans for years, and happen also to have a geological nose, you can make some pretty shrewd guesses as to where there might be an awful lot of gold. Certainly where it'd be worth sinking a borehole.'

'Roger the Klondiker. You're an adventurer at heart.'

'I'm sure as hell not staying in academic geology all my life. I want lots more fieldwork, the more the better. I'm thinking of joining this sailing clipper to investigate a fjord system in South America. Lovely stuff: a proper sailing ship. I want to see things.'

He has already seen things, including Mount Paramount in Alaska—minus the halo of stars—which any moviegoer would recognise. I remark that this is not the first time he has sounded bored with his present niche in geophysics.

'Well, of course, for plenty of geologists, including some aboard this ship, I may say, it's just a job. They could be doing anything, really. They mostly do it perfectly adequately but with no particular enthusiasm, so they bring no particular imagination to bear.'

Nobody ever seems to go up on deck just to look at the sea. They go up to inspect gear or sunbathe obsessively with a dog-eared bestseller from the ship's library, but on the whole are not to be found staring into the scud and dazzle and maybe wondering about the strangeness of being in a floating speck suspended miles above mountains, like the silver dot of an airliner over the Alps. Nor, with a few exceptions, do they seem aware of sailing across a wasteland which, even five years ago, would have teemed with dolphin and porpoise and now stretches to the horizon unbroken by anything other than the occasional flying fish. To see the world's greatest ocean suddenly empty within a few years is to be filled with a foreboding which cannot be dispelled. Something is happening below us which geophysical oceanography, its electronic gaze fixed firmly on the lumps and hollows of the seabed, is missing. Yet it needs only to come up on deck and look at the empty waves, the nearly birdless sky.

If the scientists aboard *Farnella* split, unsurprisingly, into those with imagination and those with less, odder fractures become apparent. On a tour of the ship on the first day Roger had showed me the bow thruster room. This led down to a horrible cubby hole containing cramped catwalks above bilges and the hydraulic columns of an electric motor and auxiliary propeller which could be lowered through the hull for fine manoeuvring when core samples were being drilled from the seabed. It was into this cubbyhole that, last 4 July, a British oceanographer had lured some American colleagues and closed the hatch on them, lowering gruel through a hole and placing a sizzling steak on a grid just beyond reach. After two hours he alerted a hapless seaman to odd noises coming from the bow thruster room and made himself scarce as the released scientists roared through the ship. The incident of the bow thruster room had entered the unofficial annals of oceanography even as, at the other end of *Farnella*, GLORIA was patiently mapping an unknown section of the planet's surface. Pranks and discovery.

'They take themselves so *seriously*, the Yanks,' Roger observes in what he thinks of as the confidentiality of the dark room. This Independence Day caper was recounted as a lark, not as savage. Still less was it an act which had quite precisely emphasised dependence. Aboard this ship the British are in numerical superiority and sure of themselves. The knockabout humour is such that a female scientist is required to become one of the lads, to the extent that when a woman geologist refers to her fiancé in San Diego I am momentarily as surprised as I would be if the bos'n himself had alluded to a boyfriend in Dutch Harbor: surprised not by the fact, but by the evidence of any adult life elsewhere. One evening a dozen of us are watching a ho-hum video film about a hard-boiled American cop who goes to Japan to track down the Yakuza gang member who offed his buddy. This provokes one of the British scientists into giving a commentary in 'Japanese' (high-pitched exclamations, aspirated 'h' sounds, 'Carry On Samurai' syntax and all). The strange thing is that Elly, a Japanese-American geologist, is sitting three feet away. She elects gracefully to ignore the whole thing. From time to time throughout the trip this most amiable man's hysterical Nipponese, presumably quite unconscious, can be heard now in the lab, now in the dining room (Brussels sprouts apparently being reminiscent enough of bean sprouts to set him off). Elly betrays no sign of offence except now and then shooting him a weary glance. Maybe oceanographers prefer their bonhomie ruthless. This is not the first occasion that she and several others, including this fellow, have sailed together; conceivably she has built up the necessary tolerance.

Elsewhere on *Farnella* are reminders that intellectual lives are being lived and thoughts thought. I am tackled at breakfast by Stuart, an electronics technician from Burbank who, over scrambled eggs, wants my opinion on Sacheverell Sitwell, Proust and Herman Wouk. He is also a musician. 'Dinu Lipatti. God, nobody will ever play the B flat Partita like Lipatti. Not *ever*. And what about Earl Wild? Fantastic technique. Do you know his recording of Liszt's *Don Juan* fantasy? Isn't it great? And there's an incredible track on the same disc, another Liszt fantasy. It's on a Meyerbeer opera . . . No, it's gone. I get really mad when I can't remember . . . *Robert le Diable*, that's it.' Stuart is in his mid-fifties, looks twenty years younger. He has a wife and houses in California. He is no stranger to life at sea, having spent several years in the US Navy. 'But I've got to get out of this business. I always fall foul of hierarchies. I never can get the politics right. I can't work out why people so much my junior always seem to wind up calling the shots. Maybe I'll start a business repairing all those videos people just junk because there's nobody to fix them. Easy money. The rest of the time I can read and play music.'

Strange shipboard rituals, too. The sea has always presented itself as a collection of symptoms while denying itself
the status of a disease. Every evening before supper ('tea,'
between 5 and 6 PM) the 'Oily-Boily Bar' is convened. The
name refers to oily boilersuits and means that anyone is free
to come and drink in the democratic surroundings of an engineer's cabin. That is, a scientific engineer, not a crew member. ('Them and Us?' repeats one of the cooks thoughtfully.
It seems to strike him as a pithy, even novel, concept. 'Yes.
Them and Us. I reckon you could say that.') The demarcation
is not at all snobbish. Many of both crew and scientists are
veteran shipmates of several cruises and greet each other ami-
cably when they meet in passage and companionway. I simply record that I never saw a crewman at an Oily-Boily Bar
session.

Another semi-ritual is 'Hump Day,' the cruise's halfway
point 'from where it's all downhill,' as somebody says. I had
always associated the expression with old hands of the Far
North—whalers, Yukon trappers, Arctic explorers—as meaning the moment when the endless winter reaches its shortest
day and the light at last begins to lengthen. Hints of despera-
tion, too, in 'Gangplank Trials,' when in the last week of a long
cruise the gangway is reportedly laid out flat on deck while scientists in full shore gear run along it carrying suitcases. 'It's a
rehearsal, isn't it? Helps the spirits. Tells you it's not long now.'
The echo is of the 'Channel Fever' which returning British
sailors used to contract on first glimpsing the south coast,
knowing they were still a couple of days or more out of
Liverpool or Glasgow.

Belatedly it becomes clear. Probably none of the younger
scientists on *Farnella* has been to a boarding school, in the military, or even to prison, and they lack a way of measuring time
internally. They are mainly family people, not seafarers at all.
Their business happens to be oceanography rather than metal-

lurgy. Somewhere in Surrey and San Diego are office doors with their names on them, lockers and lab smocks belonging to them, named spaces for their cars outside, a house within easy commuting distance. They are not gypsies or nomads.

The days pass, the ship goes back and forth along its lines, the shaded portion of the circle grows. No disasters, only the single hitch of the weather. The wind is stiff, the waves high, as if we were perpetually on the edge of a storm system 500 miles away. The seas hit us at an angle, nudging the bows off course. There are worries about possible damage to GLORIA's cable which thrums like a steel bar over the stern, slackening in troughs and then tautening with a snap. The scientists confer with the Captain and agree to knock a knot off the speed. This means computer work so the scans are not distorted. A new seamount is discovered whose foothills were first spotted on the previous leg while travelling in the opposite direction.

'I don't know what the hell that is,' says Mike, pointing at a gravelly looking portion of a gravelly looking picture. Doctors poring over an X-ray.

'I reckon it's a nodule field,' says Roger. 'I've seen something just like it elsewhere. Those are really sheer cliffs,' (rock faces of 3,000 feet which the world's mountaineering community would race to get their pitons into were they on dry land) 'and there's less sedimentation here so we're in the lee of the local current system. That means *this* stuff—a chewed ballpoint taps the picture—is boulders. Detritus that's fallen over the edge or sheared from the face. Doesn't look like any new activity here, so these'll be nodules, half buried by the debris and extending out to—wow, it may be off this leg too so it's a big field. Anyway, bet you anything that's what it is. I feel it in my bones.'

'And can Roger's famous bones be wrong?' murmurs Mike.

The tone of these conversations, the friendly offhand speculativeness, betrays like nothing else the distance travelled over the past one hundred years by the Earth sciences, and particularly geology. The casual references to old and recent rock formations are maybe not novel, but the confidence belongs to an era which has an unassailable timescale fixed in its collective scientific head. The seabed we are mapping is not immeasurably ancient. The area we are in, not far from the Hawaiian Chain, is volcanically active and the seabed comparatively mobile. Johnston Atoll, for instance, is a ruin of its former self, far smaller and sinking steadily. (Its reefs show it is sinking just slowly enough for coral polyps to keep pace.) Even on dry land the most monolithic features often turn out to be quite recent affairs. Not far below the summit of Mount Everest are chunks of oceanic crust with fossil sea shells embedded in them, indicating that the entire Himalayas were a stretch of seabed until India banged into Asia and the impact pushed the rocks five miles straight up into the air. That was a mere forty million years ago, in geological terms practically the other day.

Aboard *Farnella* there is easy talk of subduction zones; any disputes arise over quite specific technical theories associated with them. Subduction zones (where the edge of one tectonic plate dives down beneath the edge of its neighbour) were not suspected until plate tectonic theory proposed them, and tectonic theory itself was a direct descendant of the theory of continental drift which the German, Alfred Wegener, had brilliantly proposed in 1915. Poor Wegener! In 1928 a panel of fourteen geologists was convened to vote on his theory and only five were fully in support. Two years later he vanished in a blizzard in the middle of Greenland and his theory suffered a similar fate until the 1950s. I myself can remember a geography lesson in early 1955 when the teacher, in response to a question from a boy who must either have been very studious or else a troublemaker, suddenly shouted: 'Continental drift is *bunk!* It is *loathsome* bunk!'[1]

And so, as the geologists speculate and tap their bitten ball-points on computer printouts, I have to remind myself that not so long ago their conversation would have been pure heresy and have evoked tirades of condemnation, not only from other scientists but also from fully robed bishops. A bare hundred years before I was born William Buckland (who was both a geologist and the Dean of Westminster) wrote his *Bridgewater Treatise* in which he categorically affirmed that Noah's Flood accounted for sedimentary rocks and fossils. To watch the endless scrolls of paper emerge jerkily from printers and plotters is to feel that this is not so much mapping a new world as burying an old, which in turn makes one wonder what fresh heresies the future holds.

Meanwhile, the question of whether or not we have discovered a hitherto uncharted field of manganese nodules looks like having to remain undecided until a later cruise. The weather is now bad enough to make us abandon our course and begin a new set of legs on a different tack.

'Not so nice for us, though. It'll mean rolling instead of pitching. But kinder to the equipment.'

'What does the forecast say?'

'We *are* the weather forecast, that's the trouble. The *Farnella*'s an official weather reporting ship when she's on station, so the forecasts out here are simply based on our own data from yesterday.'

---

[1] Yet there is still dissent. Dr. Glebb Udintsev at the Moscow Institute of the Physics of the Earth consistently refuses to believe in plate tectonic theory, specifically rejecting the idea of subduction. His view is that the Earth is slowly expanding. This would indeed explain many tectonic phenomena, though much seismic data would have to be ignored or tendentiously interpreted. Unfortunately, it is not entirely easy to disprove the expanding Earth theory since any rate of change would be infinitesimally slow, too much so to be revealed by bouncing and timing radar signals off the Moon, for example.

I like the idea that we are helping invent the world's weather but can see it is unhelpful having no higher authority on which to rely. Still, the measured tones of a radio weatherman repeating one's own reports might well endow them with an official quality which would suddenly make them credible as predictions. Up on deck it is exhilarating as the ship buffets into the wind, shouldering clouds of water back over her superstructure. Another day denied the sunbathers, their towels unspread, their copies (courtesy of the Marine Society) of Tom Clancy, Len Deighton, Clive Cussler, Cruz Smith, unopened. The library contains dozens of these more or less identical Cold War what-ifs: unmemorabilia which after the past few years have suddenly come to seem like fossils, requiring of their readers a streak of the antiquarian, even of the geologist. By contrast, nothing could be more present than this ocean leaping past and over. It raises its crests to the horizon, empty and brisk, at once a locus of sublime vacancy and massive energy. When the tropic sun blazes through cracks in the scudding clouds a brilliant ultramarine floods back into the water. Simultaneously, its tusks of foam gleam white while peaks and crevasses wring colours from the constant motion. At these hot, radiant moments the sea is transformed into prodigious and arcane machinery, liquid clockwork glinting with moving parts. In the intermittent bursts of light it looks like what it is: the planet's gearbox mediating and transmitting the motive power of the sun.

In early afternoon the noise of lighter engines detaches itself from the background roar of the ship's funnel. An unmarked grey P3 Orion flies past very low off the starboard bow, its two inboard propellers feathered. Ahead, it makes a wide sweep, banking so its wings flash against the storm clouds like a fulmar's, then returns for another pass. Coming from in front, its crew will easily be able to read the great blue logo painted below *Farnella*'s bridge: a circle containing an anchor and the legend 'US Geological Survey.' This identification will be superfluous since the aircraft is one of the US Navy's long-range anti-submarine patrols and will automatically have dropped sonar buoys to learn what we are up to. They will already have heard our cacophony of pulses and identified GLORIA's correlation signal. A hand is raised behind a cockpit window as the plane drones by. It climbs away to the west trailing dark exhaust and is soon lost in the cloudbase. The ocean feels momentarily emptier for its departure.

There is a real oddness in all this watching and listening. Held firmly in the US Navy's electronic gaze as we are, the *Farnella* is herself sizzling with codes which enable us to listen to the seabed and define it, while undoubtedly provoking the attention and maybe even aggression of untold creatures below. Nobody is listening out for them at the end of this bizarre acoustic chain, however, and they will only be heard if the wavelengths of their signals happen to coincide or interfere with those of the scientists' finely tuned sensors. Certainly, no human ears are on the alert for them. In the meantime we are also listening to ourselves via our own weather reports. The ocean which surrounds and supports us manages not to impinge on our senses except to induce queasiness in one or two people.

'I'll never go on another geophysical cruise, I swear it,' Pattiann says next morning. 'They're so *boring*. A sampling cruise is much more interesting. The dredge brings up lovely great chunks of stuff. At least it's something to paw over and get your hands mucky on. All these computers—that's not my idea of geology.'

She assumes I am as bored as she, but I deny this.

'Still, next time I'd go on one of our sampling trips,' she recommends. 'Or a JOIDES drilling and coring cruise where you get to see real stuff from below the bottom. Bits of the living planet, you know. This electronic surveying's too hands-off for people like us.'

Pattiann, who must be somewhere in her forties and hence among the older of the scientists aboard, may be making a rhetorical gesture implying a companionable solidarity, or else the more radical point that scientists' differing views of the technology at their disposal might have much to do with their age. Pattiann's lineage is from the nineteenth-century oceanographers who were excited by touching and smelling and tasting whatever they studied. In fact, a certain scepticism about modern high technology, not least its prohibitive cost, is detectable in all but the youngest scientists at one time or another. Obviously, there is no cheap way to map a seabed accurately from a distance of five kilometres; yet an elusive feeling of there being a discrepancy somewhere lies on the printouts like a glaze. It is as though discovering what there is beneath the sea, learning about places on Earth which no one has ever seen, ought to be momentous, even personal. Yet it is largely automatic, done even as we sleep. Machines make the signals, listen for their return, gather navigational data, juggle it around in digitised fashion, convert the digits to a visual analogue and produce an image for a geologist to tap his teeth over. All well and good; but to requote Roger's maxim: in science, as elsewhere, you can't get something for nothing. At some point there is always a debt, a deficit, a loss. A law suggests itself, to the effect that the way any data is read is a function of the way it has been gathered.

This cruise is unusually short, a mere fortnight at sea. It has been timed so people will be able to fly home for Christmas. The mosaics of the GLORIA and air-gun prints grow on the chart table. The sweet, hot-blanket smell of popcorn drifts from the other end of the lab where a machine is anchored by Velcro strips near the swaying gravimeter. Katie stands her watch in the bay which houses the electronics for the two midships sonars. Or rather, she mostly sits. Only occasionally does she need to get up and jot some figures with a felt-tip pen on the printouts inching their way from the plotters. She clicks a switch or two, changes a stylus, checks the course, goes back to her book.

It is pleasing to think that most of the ocean bed never will be seen directly by mortal eye. In comparison to that vast area the ground which might be covered (at such prodigious expense) by manned submersibles is virtually nil. One might compare it to travelling across Asia by oil-lit hansom cab with the conditions of a Dickensian fog outside and then claiming to have seen the world. At the same time, sitting in *Farnella*'s lab and looking at the banks of instruments, computers, screens, laser printers, plotters, popcorn poppers and their umbilical harness of cables, all of it sailing steadily along a knife-edge course five kilometres above the Late Cretaceous seabed, one cannot help reflecting on *Homo*'s fierce if limited intelligence. The remorseless taxonomy of a century ago, using primitive but ingenious measuring devices as well as guns and nets and killing bottles, seems far distant now; yet even that distance is as nothing compared with the exponentially opening gap which separates *Homo* from the rest of Earth's organisms.

Katie is doing her watch early because 18 December is her birthday. Since she is known to be fond of rabbits there are celebrations tonight with rabbits as the theme. We have already signed a joint card and are encouraged to cut loose with the fancy dress. 'Rabbits, you guys. Anything goes. Raid the hold.' Not for the first time a deep current of hysteria is apparent and in these final three days of the cruise becomes a good deal closer to the surface. Katie's colleague Sue is for this trip the scientist nominally in charge of representing the US government's interests. Despite the protestations of power-sharing, democracy and cooperation, somehow or other it is Sue's judgements alone which have been prevailing this last week. She has a career and a CV to think of. A failure to complete a geophysical survey for which she was responsible will be remembered, if not held against her. Unknown to the rest of us she has taken a chance. Despite the weather and *Farnella*'s reduced speed, Sue has insisted we complete our allotted task even if it means cutting things fine.

At tea next evening the news reaches us from the bridge. There is now no way the ship can arrive back in Honolulu on the Friday morning as planned. Indeed, if the weather gets any worse we could be anything up to twenty-four hours late. Suddenly, there is a possibility that everyone may miss their flights home. Christmas Day is on the following Tuesday and the flights had anyway been hard enough to book. Over peaches and custard and the remains of Katie's birthday cake all pretence of the cool, scientific approach evaporates. Hard words are spoken about stubborn and cocksure youngsters whose inexperience jeopardises decent folks' family lives. The scientist who so recently stood her GLORIA watch, slightly the worse for drink and wearing a long pair of cardboard rabbit ears, puts down her spoonful of custard and stares at her plate, cheeks crimson. It is pure boarding school. Then the US government's geologist bursts into tears and runs from the room.

'Huh,' says a heartless Brit, 'we'd all like to be able to do that, wouldn't we? Do *we* feel any better about missing our Christmas?' And looks round rhetorically for assent. A deputation of girls plods off to the lab to comfort Sue, who has taken refuge among the charts and plotters of her disgrace. 'Right, then. I'm off to the bridge to kick ass. Anyone coming?'

Presumably each expedition becomes characterised by its own catchphrase. This cruise has acquired two, one of them written up on the lab's steel bulkhead in bar magnets, 'Yee Haa!': a cowboy's yell which flew one night out of the drunken Oily-Boily Bar. The other phrase is 'Kickin' ass,' which has recurred on surely every one of the video movies we have sat through this last fortnight—movies about cops and cops and cops, winsome black cops and shitty white ones, Marine sergeants and top gunners—so that suddenly the whole of American culture seems embodied in a catastrophic anger. It is with these furious heroes we are supposed to identify, these men with their scratched biceps and bared chests and 400-word vocabularies; so that a mild geologist from Godalming, put out because he may not make Honolulu on time and hence his flight back to England for Christmas with the family, says 'I'm off to the bridge to kick ass.'

During the next day, though, things look up. With all the equipment safely back on board the *Farnella* is able to pick up speed. The chief engineer, himself due in New Zealand for Christmas, is coaxing every last revolution out of the engines. ('All right for him,' says his junior. 'He's retiring after this trip. I'm the poor bugger's got to put in new piston rings over the holiday.')

In the event we dock at 6 PM on Friday after all and nobody misses their flight. Sue is back to being one of the boys. GLORIA is back in its cradle. The rolls of printout, the reels of computer tape which are the only tangible evidence of the invisible seabed we have been criss-crossing for the past two weeks are safely packed up. It has all been a great success, is the verdict. No equipment lost, nobody swept overboard, unlike the luckless oceanographer who had disappeared recently one stormy night in the Bristol Channel. In fact, a cushy number all round. Even the sea has a satisfied look to it as it mulls around the pilings of Honolulu harbour. It has so simply kept all the secrets it had which were worth keeping.

Cabs arrive on the quay to take the scientists on a last-minute shopping spree before their flights next morning.

'Off to Hilo Hattie's to get a really *crucial* pair of shorts,' is Roger's valediction. They vanish in a cloud of exhaust. Stuart appears at the rail next to me, slightly mournful in shore kit.

'Do you know Rosalyn Tureck's performance of the "Forty-

Eight?" What do you think? Total contrast with Gould's read-
ing, I guess. Some of his tempos seem downright crazy but
God, I remember when his first Goldberg came out in the 50s.
We'd never heard Bach playing like it. Nobody had. The energy!
Everyone thought the Goldberg was dead, academic, cerebral
stuff, you know? But it wasn't. It was *alive.*'

   He stares down at the crack of water between wharf and
hull. His voice is more animated than at any time in the past
fortnight.

II

'Nothing is more tedious than a landscape without names'

Immediately to the north of Hawaii, scattered across the Murray Fracture Zone, lie the Musicians Seamounts. They stretch for maybe 200 miles, from Strauss in the north to Mendelssohn in the south. There is a Bach Ridge and a Beethoven Ridge. There is also Mozart, a considerable mountain rising from the abyssal plain five kilometres below to within 900 metres of the surface. Mount Mozart, while a fairly minor affair by suboceanic standards, is therefore slightly taller than Mount Fuji, although of nowhere near such classic proportions.

The presence of this random clutch of composers *engloutis* in the middle of Pacific wastes is a reminder of how much of the physical world belongs in its taxonomy, description and name to the Western nations. It is also a reminder that in a sense things do not exist until they are named. Before that, everything partakes of a state of undifferentiated chaos which is never a neutral matter to human beings but carries a degree of menace. To name something is to take control of it. It could be argued that the Old Testament story of Genesis was less a matter of creation than of naming, of God taking control of chaos. Whereas before, the pre-Universe consisted of a kind of primordial babble, God-grammarian sorted out its constituent parts and uttered some solid nouns—dualities, mainly: crude oppositions such as light/dark, heaven/earth, sea/land. How he had entertained himself before this basic act of intelligence is open to speculation, but if he was anything like the humans he created (and according to Scripture he was) he was bored, repelled and finally menaced by a universe which was still a state rather than an infinite collection of objects. Ever since, *Homo* has felt the same and travellers have gone about the globe as adventurers, conquerors, sightseers, nomads and scientists, naming its parts and often bestowing on them their own proper names as well as those of their friends and sponsors. In his short story 'Colomba' (1840) Prosper Mérimée's English heroine, Lydia Nevil, takes pleasure in learning the names of places on the Corsican coast as she passes in a schooner, for 'nothing is more tedious than a landscape without names.' Many a sea captain found his spirits insupportably lowered by a coast such as that of Africa, when whole days might go by without sight of a single named feature. It would presumably have made little difference knowing the local tribespeople had their own names for the hills and capes and rivers. Being illiterate, they would have been ineligible to bestow valid names because unable to write them on a map. Only cartography can remove names from merely local usage and bring places into international being.

The desire to tame a threatening landscape by subjecting it to the control of language can be seen in the old Greek name for the notoriously treacherous Black Sea: the Euxine, or hospitable. An extension of this may result in the temporary renaming of already well-known places. In World War I when British troops were mired into the static and murderous wastelands of trench warfare, micro-maps were devised for the tiny localities which bounded their lives. London place names were wistfully bestowed on slivers of Belgian and French farmland. What a year or two earlier had been 'Quineau's acre' or 'Drowned-cow bottom' were now Haymarket and Leicester Square. This yearning domestication of threatening foreign places is a common enough trope in wartime ('Hamburger Hill') and came equally naturally to Pincher Martin, William Golding's wrecked sailor. Almost his first act on being able physically to patrol the Rockall-like Atlantic islet on which he was washed up was to give its features familiar names like Prospect Cliff, High Street and Piccadilly. This was in recognition that, unnamed, the place of his marooning would have remained inimical to him as well as invisible to rescuers, being quite literally off the map.

A Mozart Seamount does, however, seem particularly arbitrary in the subtropical latitudes around Hawaii. Odder still, it is equally close to Gluck and Puccini Seamounts, just as Haydn is to Mussorgsky and Beethoven Ridges. Clearly it is useless to look for any correlation between the physical proximity of these seabed features and the chronology of their namesakes. Somebody must have thought 'We've done poets, now let's do composers,' much as local councils name the roads of new housing estates. It is only since the invention of a technology powerful enough to map the deep seabed that the finding of names has become a pressing issue. By the early years of the twentieth century most of the planet's territorial features had been mapped and named, with the exception of the remotest hinterlands like Antarctica and the Amazon jungle. Sidescanning sonar is now revealing ever more details which for geologists, if for nobody else, need to be identifiable by name. As far as the military is concerned the situation remains equivocal. Strategic seabeds like that beneath the Arctic ice cap have been extensively mapped by NATO and Russian submarines, but their charts remain classified. There are projects for civil mapping and geological surveys of North Polar waters, but they remain projects until somebody donates a nuclear submarine to an oceanographic institute.

In order to cope with the need for new names on new charts there are various regulatory bodies which amount to a more or less official international committee on names. There is, for example, BGN/ACUF: the US Board on Geographic Names, Advisory Committee on Undersea Features. There is also the Monaco-based GEBCO: General Bathymetric Chart of the Oceans, an organisation founded at the beginning of the twentieth century. These bureaucracies are constantly turning out documents, indexes, guidelines, lists of eligible names and the like. Very occasionally a lone human voice cuts through it all, like Robert L. Fisher's in his 'Proposal for Modesty.' In this he inveighs against

> parvenu scientists who offhandedly baptize a deep-sea . . . feature that may have been known and well-explored—even if possibly unnamed—earlier, or even one bearing a long established name in another language . . . Some . . . apparently know so little about historical courtesy, significant commemoration, or even good taste that the seafloor is becoming littered, and the literature of marine geology and geophysics cluttered, with personal, in-group, self-aggrandizing, back-scratching, trite unimaginative ('14°N Fracture Zone') names or ugly acronyms ('GOFAR Fracture Zone').[1]

The Musicians Seamounts are an example of bureaucratically approved naming. It was likewise decreed that a group of submarine features off the south-west tip of Ireland should be named after Tolkien characters, which explains the Gollum Channel. The bureaucrats do not have it all their own way, however. Now and then the working names which pioneering geologists assign their discoveries stick, in all their whimsicality. A few years ago Quentin Huggett and his IOS colleagues were mapping some seabed fields of manganese nodules with GLORIA when they found a series of hills which they needed to be able to identify as they worked. One became Nod Hill, a second (felicitously named on Christmas Day) Yule. A third hill became Mango while the fourth—

[1] Robert L. Fisher, writing from the Scripps Institution of Oceanography in *Geology* (June 1987).

unfortunately never discovered—would inevitably have been Knees. Nod, Yule and Mango Hills remain to this day and probably always will, long after they have been stripped of the asset which gave them their name, like the Gold and Ivory Coasts.

A more famous and no less whimsical example is of an area of Atlantic seabed to the west of Spain which celebrates British biscuits. This centres around the Peake Deep, modestly named after himself by the ship's captain who discovered it. A later expedition from Cambridge found a long, shallow depression in the same area which they loyally named King's Trough. Then they discovered a second deep near Peake Deep and called it Freane [*sic*] Deep. Further surveying disclosed two ridges between these features which became respectively Huntley and Palmer Ridges. Finally, the trip was completed with the identification of Crumb Seamount.

In the late 1980s, while mapping the 200-mile EEZ around Alaska, GLORIA at the end of one of its turns revealed an unknown volcano beneath Soviet waters. Quentin Huggett, interested in pre-Soviet Russian anarchist movements, reported its existence to the Soviet Academy of Sciences with the customary apology for unintentionally having 'spied' into Soviet waters and suggested it should be called Kropotkin Seamount in honour of Peter Kropotkin, the celebrated geologist and anarchist. It so happened that Kropotkin's nephew, himself a geologist, was on the panel of Academicians considering the suggestion and reportedly the proposal raised a laugh in the relaxed climate of *perestroika*. A certain edge to the laughter might have come from the knowledge that Peter Kropotkin had been distantly related to the Romanovs and his nephew is considered by some today to be the Russian citizen nearest in succession to the Russian throne. In this particular case, and beneath the international exchange of jocularities, curious games are perhaps being played. For while the impli-

cation of the story—heard from the Western side—is that scientists with superior technology could tell a country things it
does not know about its own territory and would happily do so
for a price, the story from the Russian side might be quite different. With military security it is never quite certain what is
known. The Bering Strait must after all be one of the areas
most familiar to Cold War submariners, and it would seem
likely that this seamount was already known to the Russians,
who might have declined to submit it themselves for international naming in order to disguise the extent of their own
knowledge.

III

MARGINALIA:
Zetetics

This chapter has constantly invoked the notion of a globe, an oblate spheroid, to represent the planet. This is reasonable, given that not only do Newtonian physics and mechanics define it as such but it looks like one when viewed from space. There has long been—and still exists—a pseudo-science called zetetics which maintains that the Earth is in fact flat. Its adherents are popularly known as 'Flat-Earthers,' a term of disparagement which connotes either stupidity or else wishful, head-in-the-sand archaism. Nor is this scorn unreasonable, given the crackpot tone in which their case is usually advanced.

When zetetics most earnestly offers its evidence, the classical procedure is for it to cite a list of mathematical and other 'conundrums' which might be taken as casting doubt on Copernican theory. It first proposes a model of the Earth which is a vast disc, an irregular plane of unspecified thickness and circumference at the centre of the universe, above whose surface the sun and stars circle on concentric paths. Its circumference is indeterminable because the edge of the known Earth is surrounded by a barrier of ice (which others might call the Arctic and Antarctic) beyond which 'the natural world is lost to human perception. How far the ice extends; how it terminates; and what exists beyond it, are questions to which no present human experience can reply.'[1] The

---

[1] In the days before flight and satellites this fancy was perhaps slightly less stupid than it appears today. Many a nineteenth-century Arctic explorer was fooled by a *fata morgana*, a complex mirage that can give an illusion of distant mountain ramparts.

words are those of 'Parallax,' the second edition of whose *Zetetic Astronomy* was published in London in 1873. There is reason for thinking he might also be the S. B. Rowbotham who published a book of the same title in 1849. This earlier date has some significance because it falls in a period of great interest and debate about the age of the Earth (see Marginalia to Chapter 5) and it is not at all surprising to find that 'Parallax' is a firm believer in 4004 bc as the date of Creation.

He begins by describing several experiments with flags, poles and ships to prove that the surface of water—and therefore also the sea—is not convex, and in due course reaches his cannon test. His argument is that if Earth were a rotating sphere a cannonball fired vertically into the air could not possibly fall back on top of the cannon. There is an engaging, schoolboy quality to this idea, like that of a child who imagines himself falling in a broken lift but able at the last moment to save himself (unlike everyone else in the lift) by craftily giving a little jump just before it crashes at the bottom of the shaft. At any rate, the conundrum of the cannonball exercised the minds of some quite elderly schoolboys during the Crimean War, and on 20 December 1857 the Prime Minister, Lord Palmerston, wrote to his Secretary for War, Lord Panmure, to clear up a few niggling points about British gunnery. He was assuming a cannonball fired in the air would not follow exactly the rotation of the Earth's surface but would to some extent be left behind. That being so, he wondered if the Secretary for War realised that the tactics of modern warfare ought perhaps to be altered to take into account the obvious fact that the range of guns must vary according to the direction they were pointed in. Clearly, if they were fired eastwards in the direction of the globe's rotation the balls would 'fly less far upon the Earth's surface than a ball fired due west'.[1]

[1] 'Parallax,' *op.* cit.

'Parallax' re-examined the accounts of the voyages of oceanographers and explorers like Maury and Sir John Ross to show where their navigation had been at fault. It was hardly surprising that a man like Ross, even though of the highest personal integrity, had been deluded into thinking he had spent four years completing a circumnavigation of the globe when all he had done was sail 69,000 miles around the inside of the Great Ice Barrier. There was an urgent need to revise the whole science of navigation, particularly knowledge of the tides, sunrise and sunset, the seasons and the laws of perspective. Some of the writer's own conclusions about the heavenly bodies were indeed radical. The sun is 'considerably less than 700 statute miles above the earth,' and 'all the visible luminaries in the firmament are contained within a vertical distance of 1,000 statute miles.' Moreover, the Moon is transparent. 'We are often able to see through the dark side of the moon's body the light on the other side.'

This is most inventive, and 'Parallax' had many disciples from that day on, some of whom cribbed his examples. In 1940 a certain E. L. Venter of Bloemfontein published *100 Proofs that Earth Is Not a 'Globe'* and also considered the cannon test, concluding 'the ball always falls back on the cannon.' (Had he conducted his own experiments, one wonders, wearing a tin hat on a private range out in the veldt?) 'That test proves that the earth is stationary. It is our proof no. 45 that the earth is not a globe.' To this he adds the evidence of the 'shadow.' 'In the tropics a six foot man has no shadow at noon and for 16 miles on each side of him men have no shadow at noon, but men farther away begin to have shadows. This test of the vertical rays of the sun indicates that the diameter of the sun is only 32 miles . . .'

This sort of thing is not like the position of men like Dr. Udintsev, the lone sceptic who doubts tectonic theory. But the point at which a serious scientist becomes a 'nut' is not without interest. The original zetetics were followers of Pyrrho, the founder of the sceptic philosophy, and their name derives from the Greek verb 'to seek.' They thought of themselves as searchers and enquirers, not believers. The example of 'Parallax' and others like him shows how swiftly a text or collection of assertions turns into a doctrine which itself becomes a comfortable hive for every bee in every bonnet. Yet there ought to be an aspect of zetetics which confines itself to an absolute scepticism both playful and imaginative. There was an admirable movement which resolutely refused to believe in the US moon landing of 1969, maintaining that the entire thing had been a brilliant hoax designed to cow and discomfit the USSR as well as to ensure Congressional votes for the further unlimited funding of NASA. The idea was that the astronauts had never left the launch pad, that the extravaganza had been created in the studio using simulation techniques and special effects and fed into millions of TV sets around the world where it was received by a credulous audience. The merit of this argument, apart from being funny, was that the average viewer could not refute it. It made the point that a TV audience will believe anything they see provided it is served up with the right trappings and couched in the approved 'History in the Making' rhetoric.

It is not only from an anarchic streak that one would nearly prefer this version of events, but mainly because scepticism constantly frees thought yet can coexist with knowledge. Only at irritable moments aboard *Farnella* would I have been tempted to argue that GLORIA was an expensive hoax and the seabed at an infinite depth. Even the denial of a true idea creates a space which vibrates with possibility.

This place has no name . . . Nor does the lost swimmer even have a place, buoyed up as he is in an illimitable steep of fluid and, for all he knows, borne along by a current. In all directions is void, whether air or water, though busy with sunlight and spangles and small events. The sea itself is calm. He wishes there were a higher swell so he could more easily keep up the hope that his boat, even though not many yards away, remains hidden by conspiracies of wavelets. He knows exactly what it would be like to be in an aeroplane flying above where he is now. He knows the burnished pane of ocean with its frozen wrinkles crossed by the aircraft's shadow. He knows, too, how words like 'millpond' only ever come into the mind when gazing disembodiedly out of a window at 20,000 feet. This leaves the swimmer with an echo from which to build a name, 'Despond,' for this locus in which he is adrift before he abandons it as hackneyed and unhelpful.

At last he works out that this place can have no name other than his own. Nothing if not isolate, he is himself an island. By mischance or gross carelessness he has become marooned on himself. This perception has a point in its favour. It is an island with room for only one castaway. In the almost impossible event of anybody else reaching its shores they would at least be coming as rescuers.

2

Islands and Boundaries

A detail from Abraham Ortelius's map of the world (1570) showing a selection of North Atlantic islands, both imaginary and misplaced.

I

Tiwarik lies little more than half a mile off the coast, immediately opposite the fishing village of Sabay which from the seaward side appears as a straggle of huts on stilts half lost among the coconut groves. In front of them on the stony beach boats are drawn up, each of which in time becomes identifiable so it is possible on any particular day to read the beach and know who is doing what. The island is uninhabited (uninhabitable, practically, since there is no water) and tiny, being about a quarter of a mile across. But its size on a map—and I have never seen a map large-scale enough to mark it—would be deceptive, for it rises to a peak off in one corner which cannot be less than 400 feet above the sea. There are no beaches, merely that one shifting coral strand maybe a hundred yards long and facing the mainland. The rest of Tiwarik rises from gurgling boulders more or less vertically up volcanic cliffs of black rock. From one quarter there is a steep sweep of coarse tall *cogon* grass up to the forest which caps the peak. Seen from the strait on a breezy day the sunlight goes running up and up through this wild grassfield. It is the same effect as with young hair and similarly afflicts me with deepest melancholy, affection and pleasure.[1]

This was how I depicted an uninhabited island where I lived for a while. I wanted to account for its centrality to my

[1] James Hamilton-Paterson, *Playing with Water* (1987).

life, while describing the lives of the villagers on the other side of the narrow strait, their habits of dynamite fishing and our nocturnal spear-gunning forays among the reefs and boulders. I noted some of the salient features of tropical offshore islets in South-East Asia: their considerable discomforts, their frequent lack of drinking water, their distance from supplies of nearly everything other than brine, rocks, harsh coral sand, thorns and greenery.

Perhaps I wrote too much about beauty—at least, about the unidyllic way the island struck me as beautiful—since really I wanted to make the point that in the terms of holiday brochures it was nothing out of the ordinary and near to nowhere very special. I observed that it was in no sense a 'paradise isle' as the tourist industry understands it, advancing its complete lack of fresh water and, more especially, its lack of a permanent beach of soft white sand as reasons for asserting that at least 'Tiwarik' would never suffer the indignity of being turned into a resort. Its unyielding, basaltic indifference to any melting aspect, to the dreamy topos (not a coco palm anywhere) would guarantee its own rugged persistence. Finally, to protect myself as much as the island, I gave it a fictitious name and was carefully vague about the Philippine province off which it lay.

Many months after I had finished writing the book wisps of rumour reached me way up the coast. Rumours in those parts being what cocaine is to Hollywood, I attached no importance to them. Then in due time I went back to the little fishing hamlet of 'Sabay' from whose shore one can see the island a mile or so away on the other side of a strait of tearing currents.

To be the biographer of a place or person can insert a murky distance between them and oneself, especially after publication. What has been a private, even obsessive, project turns overnight into an implied claim to special knowledge or scholarship when really all one had on one's mind was love and curiosity. It becomes easy to retreat into snappish inward pronouncement that while anybody may be a greater expert on the subject, nobody else has quite the same affectionate eye . . . and up drifts the murk. But on this occasion I sat on an empty oxygen cylinder on 'Sabay' beach and gazed through a clear lens of air at 'Tiwarik,' picking out the details which I still felt had written me as much as I them. On that very outcropping I had lived, had fought a grass fire, dried my catches, had worms, been bitten by centipedes, had watched a pair of sea eagles come and go to their nest in the cap of jungle with fish in their claws. I had glimpsed much else besides, and often felt I had voyaged on it further than the island's small boundaries.

'No more,' said a friend. 'It's been sold.'

A knifelike stroke. Studiedly offhand, though: 'Oh? To whom?'

'Japanese. Very rich. *Ayy* . . . very big project, James. Very big plans.'

'For *that*? Oh, nonsense, *Baka tsismis lang.*' Just another of those rumours (narcotic, stimulant, currency). Just the favourite Filipino pastime of telling tales of projects which are going to transform the hardscrabble of living, tales which occupy a psychic territory as much as a local site. Sunken galleons . . . Japanese war chests . . . Chinese pirate hoards . . . wood-burning power stations . . . The tales fade, re-emerge, fade again. The captains and the scuba divers depart; the phrase 'feasibility study' drops from conversations. Things go on being the same.

Not this time, apparently. This time they really were to change. I paddled across to 'Tiwarik,' half despondent and half thinking there was probably nothing to worry about. The familiar difficult strand, the familiar steep climb up the cliff path were the same. But up on what I knew privately as the Field of Crabs were signs that people other than local farmers and fishermen had been there. Ominous pegs had been hammered into the baked soil. A shallow pit had been dug and a red-and-white surveyor's ranging rod stuck up out of it like a thermometer in the mouth of a sick man. Only in the most oblique and imaginative sense had I ever thought of 'Tiwarik' as my island. Now it wore the anonymous, severed look of real estate.

For an hour or two I wandered about without being able to reclaim it. I stood in favourite places and looked down at the suck and surge of water, down through water to the island's roots. Those boulders, ledges, shelves, coral palaces; those blue thoroughfares and weedy balconies: I knew them, had examined and hunted every inch of them by day and by torch-light. While their aerial map was still familiar they themselves had withdrawn, taking with them whatever it is that makes places vibrate when looked at with a certain eye. Over the horizon comes the world; the eye in distraction flickers and clouds and at once even the rocks shrink in upon themselves like touched anemones. The land reverts to a blob.

So I left. Over the next year or so stories reached me out of which I built my own fretful picture of the vanishing of 'Tiwarik' and the creation of the 'Fantasy Elephant Club,' as the new resort was apparently to be known. The lack of beach would be no hindrance to the wealthy Japanese visitors who would be flown there direct from Manila by seaplane and helicopter. The last thing they would want to do was swim. There in the chalets where the Field of Crabs had once been they would live in the pallor of circular neon lights and beneath these coldly fizzing haloes be massaged by geishas, recline on vinyl couches to watch porno videos, gamble and sip Chivas Regal and other drinks peculiar to duty-free life . . . I hoped the fish eagles had lifted off disdainfully at the first roar of the chainsaws. Where they had nested a cement tower was apparently now going up. I envisaged an oriental folly, a stylised ninja pagoda symbolising the martial art of Third World property development. I kept hoping that some of 'Tiwarik''s old, implacable quality would assert itself: plagues of centipedes, perhaps; the belated discovery that the Japanese military had tested anthrax bombs there in 1944 and that all the soil more than a metre deep was virulent; even a Krakatoa-like eruption of the volcano opposite which had not let out so much as a squeak of steam in the past 50,000 years . . . Then I heard that three people had been killed—labourers, possibly, but not from 'Sabay.' I packed my bag and went back to the village to see old friends and separate news from rumour.

This time there was a choice of two oxygen cylinders on the beach on which to sit. Several families depended for their income on catching small, brightly coloured species of coral fish, bagging them in plastic with a litre or two of seawater and a squirt of oxygen and shipping them off to Manila for export to the pet shops of Japan and the West. (The mortality rate was probably 80–5 per cent, but as long as they left 'Sabay' alive it was not the fishermen's problem.) Now from these empty cylinders the view across the strait was novel indeed: of an unfamiliarly shaped island girt with rows of what seemed to be white cement bungalows with red roofs. What had once been an empty stretch of water was full of small craft ferrying groups of labourers and materials. Even as I sat, a *bangka* carrying a large cylindrical water tank crawled heavily out of the difficult shallow channel off 'Sabay' beach, its outriggers ploughing the water instead of skimming its surface. Elsewhere along the beach were long heaps of sandbags. Clearly, two of the disadvantages I had wishfully imagined would guarantee the island's immunity to change were being remedied almost with disdain. No drinking water? Fetch it over in huge tankfuls. No beach? Take 'Sabay''s across in sacks. What, then, is an island? The author of *The Island Within* surveys his kneecaps sticking up out of the water in his bathtub and considers them very much part

of his personal mainland.[1] The image has a geological apt-
ness. 'Tiwarik' is as much a part of the mainland as 'Sabay' is.
It just happens to look like an island because the land
between was low-lying enough to have been invaded by the
sea. Its flora and fauna are scarcely affected by the intervening strait. The weather is that of the mainland, birds and
seeds fly to and fro. From time to time people had made
efforts to cultivate small patches of its total thirteen hectares,
though lately this amounted to little more than occasionally
cutting the *cogon* for thatch. The island's crown of virgin jungle is a miniature version of those vestiges still surviving in
gullies and ravines high up Mount Malindig opposite.

'Tiwarik,' then, is a crumb fallen from the mainland, made
of the same dough and nourishing the same plants and animals.
At the time I built my first hut on it the island had no economic
function of its own. Yet it did form a casual part of several
economies. Locals fished there and, especially when caught by
sudden squalls or currents, would hole up on it until conditions improved. On ordinary days they might land and build a
driftwood fire on the coral strand, toasting a fish for lunch. It
was also used by travellers. The archipelago is full of migrants
undertaking long and dangerous journeys in frail craft with
ropy engines. Some of these travellers are landlubbers appre-
hensively trying as cheaply as possible to reach a city like Cavite
or Batangas or Manila where they have heard jobs are to be
had. But most are born boat people who give the impression of
being refugees from dry land. Visayan fishermen spend weeks
away from home, drifting from province to province, from one
favourite fishing ground to another, catching and selling. Some
of them claim to have no particular home but, gypsy-like, roam
these central seas often with only a language, a dialect and a
place of birth to give them geographical identity. Any of them

[1] Richard K. Nelson, *The Island Within* (1990).

at any time may haul their boats on to 'Tiwarik''s little shore for a few hours to mend nets, cook a meal, calculate how much rice and fuel they will need to buy across the strait before pressing on again. Still other maritime vagrants are smugglers and pirates. Why else would a big, thirty-foot *bangka* from Romblon be carrying at least six boxes of grenades and a .50 calibre machine-gun hidden beneath a nylon sail? They were the most affable of all, catching me mending a plywood flipper. I gave them cooking oil and a disposable butane lighter and in return they offered me a grenade for fishing. When I declined, saying I would rely on my speargun, they said 'For self-defence, then. There are a lot of bad characters around these parts. Us, for example.' We all laughed, I a little uneasily; after an hour or two they left, waving.

Since 'Tiwarik' had no population and nothing to offer except a bit of dry land in the middle of a lot of water it was on nobody's itinerary and was no one's port of call. It lay at the crossroads of no particular routes, it formed no conceivable milestone in anyone's journey. Yet it was there to be used, to provide refuge or shelter, sticks for fires, corals and boulders for fishing. Or, for the reflective, it offered a place where one could hear only the sea's rinsing murmur, the cries of birds and, at night, the tiny hollow sigh of a lamp wick in its glass chimney.

Now this place no longer exists, and I need no reminding as our *bangka* noses on to the Fantasy Elephant Club's new beach. The tangle of boulders and thorns which had always hidden the foot of the cliff is gone. In its place is a concrete sea wall which at one end abuts the foundations for a small pier. Grey cement teeth stick up out of the blue water sprouting tufts of rusty reinforcing rods. On the spot where I had pitched a flapping shelter during a storm on my first visit to 'Tiwarik' many years before stands an octagonal, open-sided beach house with at its centre the beginnings of a circular bar surrounded by polished marble stools. At empty stone tables sit a variety of site officials—architects and engineers—waiting for a boat, as well as a blue-uniformed guard with a pistol and a walkie-talkie. No, he says, it isn't possible for me just to stroll on up and look around. This is a Japanese operation and things are run in an efficient and security-conscious manner. Why do I want to visit the island, anyway? I explain that I had once spent time here, had lived here alone, am curious to revisit it. The guard calls up, is told to wait. I sit down on a marble stool and watch relays of sweating boys stagger beneath the sacks of sand they are unloading from a *bangka*. Time passes; the guard speaks, his radio crackles back; more time passes. Finally, I and my two friends from 'Sabay' are allowed to walk up to the site but are reminded that when it is finished the Fantasy Elephant Club will be exclusively for Japanese members.

The precipitous footpath is gone. To replace it a steeply curving road has been bulldozed across the face of the cliff. As we walk up we are passed by a roaring truck full of cement and trailing sooty fumes. At that moment the last vestiges of 'Tiwarik' vanish in clouds of carbon. We stop on a curve where labourers are digging a trench for a power cable. They are from 'Sabay' and I fished with one of them four years before. He tells me this is the site of the accident a month or so back, just before Christmas. Some boulders fell out of the freshly cut embankment and crushed three workers from up the coast. Two died on the spot, a third is in hospital in Manila and likely to die. I hope their families have been compensated. Oh yes, says my informant's workmate, they were each given 30,000 pesos (a little more than $600).

Once at the top I am unable to recognise nearly all the familiar landmarks. I cannot even be certain where I saw the pegs and surveyor's ranging rod on my last visit. Much of the Field of Crabs is now landscaped and has been disguised as a miniature golf course. A tennis court is even now being sur-faced and a small swimming pool receiving a first coat of oblig-

atory blue. Individual bungalows—double glazed, air condi-
tioned, self-contained—are disposed among tasteful arrange-
ments of rocks already planted with flowering shrubs. The centrepiece of the development ('the masterpiece,' as the assistant
to the architect told me down below in the beach pavilion
within the latter's earshot) is the clubhouse itself, a long, low,
white palace whose foyer is painstakingly decorated with
appliqué designs of shells set in cement. This is apparently to
contain all manner of restaurants, *sushi* bars, guestrooms and
steam baths. At the moment it holds a good few Filipino
labourers wearing the snipped-off corners of plaster bags on
their heads as sunhats. A siren sounds from up in the forest, a
strident wail which sends an instant image of escaped prison-
ers fleeing through the mind, but it is only to mark the beginning of the labourers' mid-morning break. Most unwrap their
merienda from scraps of newspaper where they stand. A few
head up towards a straggle of huts pitched against the steep,
rocky slope in the tattered scrub which the lower reaches of
the forest have now become. Even within the forest itself there
are signs that trails have been hacked to the top. I can see no
pagoda but my companions tell me that a chalet is planned for
'those who want to be alone.' What, then, of the pair of eagles,
my familiars? I had tried not very hard to identify them, flicking through bird books for South-East Asia, but each time I
visualised them the print slipped from their wings as they
soared away, unnamed. Assuredly they were gone. In one corner of their former home a level patch has been gouged and on
this now stand three immense concrete legs.

'Those are for the electricity,' my companion explains. 'The
cables will come across from the mainland.'

And so, without need for bridges and causeways ('the
whole point is that it is an *island,*' a Japanese will tell me stiffy
some hours later), the Fantasy Elephant Club is to be firmly
tethered to the mainland by umbilical cords. The great weight

of unsupported lengths of high-tension lines explains the size of the pylons, which will have to withstand the stress of typhoon winds. It is possible to make out three similar pylons tucked away among the shrubs on the distant promontory. Until the project was hatched, this particular corner of the province must have been well down on all priority lists for a surfaced road and electrification. Now, thanks to foreign developers, new poles have sprouted along the dirt track on the mainland and 'Betamax' film shows are daily and nightly entertainment where before the villagers played cards and gossiped cruelly by candlelight. 'Progress,' as the mayoress of the municipality is fond of saying. Yet there is little risk that someone who knows the area will melt unresistingly into admiration for the foreigners' altruism. It is the province's impoverished power company which is meeting the bills for bringing electricity to the island. And over in 'Sabay,' where it is reasonable to suppose the developers might in self-interest have improved the villagers' inadequate water supply, I noticed the new hand-pump had been installed by local Rotarians, just as the recent hard surfacing of the sandy lot used for basketball had been paid for by the Lions Club.

Something had gone sour at 'Sabay,' I discovered, though I never found out precisely what. The villagers disdainfully said the Japanese were arrogant and too mean to work for at forty pesos a day (eighty-five cents). They claimed they could earn far more from their usual fishing activities. It was only the other villagers up the coast who were too lazy and ignorant to fish who would consider forty pesos a possible wage . . . There had been friction right from the early stages of construction, I gathered. At any rate, the relations of 'Sabay' to this new world on its doorstep were decidedly *malabo*: cloudy, murky, ill-visaged. It seemed the neighbouring village was to do better in terms of the newcomers' patronage since the developers were indeed sinking a borehole there for the island's drinking water. Later, I found out something else which might explain this state of affairs, since I never seriously believed my friends' protestations (no doubt cued to my own responses and facial expressions) that the island had been despoiled, their traditional lives of picturesque poverty ruined for ever by video recorders and concrete pylons.

It turned out that what was taking shape on the island was only 'Phase One' of a far larger project. 'Phase Two' was destined for the mainland: 252 hectares which, if all the requisite planning permissions were hustled through, would comprise an eighteen-hole golf course, a clubhouse, cottages for guests, a two-storeyed home for the (Japanese) aged, a private airstrip/helipad, a skating rink (presumably roller rather than ice), a shooting range and a coffee shop. There was also talk of a casino, said the mayoress, although that might eventually more prudently be sited on the island. Further facilities to be offered would include windsurfing, sky diving, water skiing, scuba diving and riding. This 'Phase Two' was intended to be open to non-Japanese visitors.

Since under Philippine law no land can be owned outright by a foreigner, the developers had to have a majority Filipino stakeholding. This was no problem for the corporation's president since he was married to a Filipina. Most of the land he needed to acquire for 'Phase Two' belonged to the village next to 'Sabay'.' In 'Sabay' itself the wife was on the point of acquiring—or had already acquired—thirty-three hectares of land for 'family use.' The 'Sabayans'' sense of grievance might therefore be partly explained by straightforward annoyance that they were not as well placed as their neighbours to cash in on this land bonanza. They clearly did prefer development to raw nature, but in order to be fully reconciled would have liked a bigger slice. The fact that all the land was at present classified as 'agricultural,' and that to change this classification required a local referendum as well as action at provincial government level, was a minor technicality. So was President Cory

Aquino's foot-dragging land-reform policy under whose meas-
ures several landless tenant farmers in 'Sabay' would have
stood to gain a little agricultural land of their own. The whole
development project was so grand and so powerfully backed
that no landless 'Sabayan' was likely to risk his neck by making
a serious protest, still less bring any sort of legal action.

So that day in 'Sabay' I take 'Phase Two' to be pretty much
a foregone conclusion. I also assume the villagers' fears about
organised crime in the shape of the Yakuza and insidious disease in the shape of AIDS are not necessarily going to be
allayed by the development corporation's president promising
'to screen all his visitors most carefully for undersirable connections' (the reference was apparently to the Yakuza), nor by
their lady mayor's assurance that she would 'eradicate illegal
activities such as prostitution' if they occurred. Several of the
more travelled 'Sabay' fishermen had seen at first hand the
results of the turning of Puerto Galera in Mindoro into an
international resort and had witnessed once-lovely Boracay
Island in Panay 'go to the devil' and require thirteen full-time
tourist security police. 'There is a price for everything,' a local
official wearing a huge fraternity ring announced sagely.
'Fighting, prostitution, AIDS, Yakuza . . . They're a regrettable
but inevitable part of the development of our country.'

I soon lose heart on 'Tiwarik' and cut short my unofficial
visit. The last thing I notice is that the tree on top of the cliff
from which Intoy and I had hung the highest, most dangerous
and most beautiful swing in the world had been cut down and
replaced with a concrete bench moulded to resemble half a
rough-hewn trunk. As we start back down the new road I discover the price of the island was 200,000 pesos. 'Tiwarik' was
knocked down for $4,300. I also hear its Japanese visitors
would be paying $240 a day for their accommodation, not
including food and the use of facilities. At the scene of the
recent fatal landslip I catch myself doing the sort of pointless

sums which nevertheless insist on being done: three days' lodging equals one life.

All these details about 'Phase Two' and the land on which it is to take shape are not by the way. They are central to the manner of 'Tiwarik''s demise and how it has lost its existence as an island. Even as, one thousand miles away, Japanese technology struggles to preserve the single eroding islet of Okinotorishima so it may qualify for its own EEZ, the same technology has here de-created an island and is building in its place an exclusive peninsula. What with private helicopters and hydrofoils and high-tension cables, the narrow strait will have even less of an isolating effect than a six-lane highway has on an urban community. The island and the mainland together form one single project. Splitting it into two phases is merely an engineering and procedural convenience and has to do with two timescales, not two separate places. By annexing—perfectly legally—a piece of territory in another country and turning it into pseudo-Japan, Japan is in effect exporting its boundary. It is mere chance that the project, and the money, happen this time not to be Swiss or American or Chinese or German. The end result is the unislanding of 'Tiwarik'.

'Tiwarik' is a good example of an island which has been lost to a kind of gerrymandering, to a redrawing of its cultural boundaries. Its mistake was to have been too amorphous, too anonymous and unclaimed except by a nomadic European who briefly thrust his own identifications on to it and wanted nothing but silence in return. That was selfish, no doubt. Also harmless. It would be a historical solecism to look back a century or two and think yearningly of the unpopulated, unneeded islands then littering the world's seas. Two centuries ago one's mind would have been quite different, with different wants and expectations in a completely dissimilar world. So, at least, we suppose. Yet islands have always exercised a fascina-

tion and—unlike deserts, for instance—are enough repositories of fantasy to be slightly chimerical. One seldom looks at an
island without also imagining it disappearing behind a bank of
fog or storm clouds which at length clear to reveal an empty
ocean. It would not be a surprise and, like a dream, one might
not even miss it. Until remarkably recently the North Atlantic
was full of islands which have now disappeared. They constituted another type of vanishing island, one whose loss is due to
the redrawing of maps, to improved cartography and navigation, but also to changed expectations.

These places hovered on the boundaries of the actual and
the credible for hundreds of years. They cannot have emerged
from nothing and neither have they entirely vanished yet in all
their mysterious aspects. The most famous of them were
Antillia (or Seven Cities), Brasil, St.. Brandan, Buss and
Mayda. Seven Cities Island had grown out of a legend of
Christian refugees, among them several bishops, fleeing the
Iberian peninsula before the Moorish invasion in 711 and
fetching up in a safe land somewhere well away in the far
Atlantic. Travellers who claimed to have visited it had found
plentiful church services and brought back sand which was
one-third pure gold dust. Not surprisingly it remained a place
of considerable interest to navigators. So did Brasil Island,
which had appeared on Dallorto's famous map of 1325 as a
large, circular island some way off Munster in Ireland. In 1498
the Spanish ambassador in London reported home that 'The
people of Bristol have, for the last seven years, sent out every
year two, three and four caravels [light ships] in search of the
islands of Brasil and the Seven Cities.'

Seven Cities was on Desceliers's map of 1546 at a position
which today would be between 500 and 600 miles off New
York. Since Columbus's day, though, mariners had been sailing
the western Atlantic with increasing frequency and by the late
sixteenth century the island had begun to shrink, moving

southwards and out into the mid-Atlantic wastes. A century later it vanished entirely. Brasil Island lasted much longer—embarrassingly so, since according to one source it 'persisted in the mind of the British Admiralty until the second half of the Nineteenth century'.[1]

The Isle of St. Brandan was even more illustrious since it could trace its origins back to ancient Greece and Celtic mythology, both of which had stories about an Island of the Blessed. Pagan myth was converted into Christian parable by putting the sixth-century Abbot of Clonfert into the role of a lone seafarer who, committing his body to a leaky coracle and his soul to God, overcomes all manner of dangers and hardship to arrive at last in a land of paradise. It was this island which Charles Kingsley chose to be the home of the water babies, using it in a curious way. He preserved the fairy-tale aspects. 'On still clear summer evenings, when the sun sinks down into the sea, among golden cloud-capes and cloud-islands, and locks and friths of azure sky, the sailors fancy that they see, away to westward, St. Brandan's fairy isle.'[2] The religious connotations were still there, too, for 'when Tom got there, he found that the Isle stood all on pillars, and that its roots were full of caves,' which implied a place modelled securely on Gothic church architecture. Yet Kingsley then invoked paganism by identifying St. Brandan's with Atlantis, the lost land of higher knowledge which to this day is still assiduously mapped by its faithful.

'The Sunken Island of Buss' was different in that its discovery was definitely recorded. It was found in 1578 during Martin Frobisher's third and last attempt to find a north-west passage. One of the smallest of his ships, the *Emmanuel*, was a buss (a herring smack of about sixty tons) out of Bridgwater in Somerset. During a severe storm off Greenland it became sep-

[1] Yi-fu Tuan, *Topophilia* (1974).
[2] Charles Kingsley, *The Water Babies* (186).

arated from the rest of the expedition. Completely lost somewhere in the region of Bear Sound, it began sailing south-eastward in a vaguely homeward direction whereupon, in the words of a contemporary but second-hand account, 'they discovered a great island in the latitude of fifty-seven degrees and a half, which was never yet found before, and sailed three days along the coast, the land seeming to be fruiteful, full of woods, and a champaign country.'[1] Almost at once the Island of Buss began appearing on charts. In the seventeenth century it was large and usually about 570 miles due west of Rockall. By the eighteenth century it had begun to shrink, improved navigation having failed to find it. Owing to its definitive position as well as to the documented date and circumstances of its discovery, Buss was never free to adopt the other islands' strategy of scooting elusively about the ocean. Instead it shrank by sinking, for van Keulen's 1745 chart observes, 'The submerged land of Buss is nowadays nothing but surf a quarter of a mile long with rough sea.' Later that century and into the beginning of the nineteenth several ships searched for Buss by taking soundings but the results were inconclusive. Probably its last appearance on a map was in the 1858 edition of Keith Johnston's *Physical Atlas*, where it is a speck in the North Atlantic.

What did the *Emmanuel* find in 1578? After Buss had finally gone cartographers were left with four possibilities. Either it was a fraud from the start or else the disoriented sailors had mistaken a fog bank or ice field for land. Or was it a genuine island which had soon submerged? The fourth and most likely explanation is that the little buss from Bridgwater had glimpsed a stretch of the Greenland coast, probably in the region of Cape Farewell. A certain amount of imagination and a vivid non-eyewitness account (Greenland is not an obviously 'fruiteful' and well-wooded country) did the rest.

[1] E. J. Payne, ed., *Voyages of the Elizabethan Seamen to America* (Oxford, 1893) p. 183.

Lastly, the Island of Mayda survived on maps longest of all. It was usually crescent in shape and placed out in the Atlantic to the south-west of Ireland, more or less where the Porcupine Seabight is on modern charts. One of the variants of its name, Asmaida, probably points to its having originated with medi-aeval Arab navigators. Like Brasil Island it began moving steadily westward across the Atlantic, in 1566 fetching up in Newfoundland waters. By 1814 Mayda had drifted much fur-ther south, on a level with the West Indies. According to one source,[1] its final appearance was on a Rand McNally map of 1906. It is anyone's guess as to what Mayda's objective correl-ative was: possibly Bermuda or Cape Cod or Cape Breton. It was most likely to have been dimly perceived America.

The modern world with GPS at its disposal, to say nothing of satellite pictures, may view these elusive islands of the North Atlantic with a degree of indulgent patronage. Yet for Mayda to have existed on maps from 1400 to 1906—albeit in a variety of locations—is a feat which testifies to something other than mere navigational error. It clearly served a real function. In its slow and stately disappearance over some five hundred years the Island of Mayda—which no one had ever seen twice, still less landed on—in effect demonstrated that something cannot come into being without displacing something else. Mayda had to be pushed gradually aside in order to allow the eastern seaboard of America to come into cartographical existence.

We who carry in our heads an image of the Earth seen from space and have noticed on the surface of its blue ball conti-nents and land masses looking identical to the ones in atlases cannot picture the world as it was to someone half a millenni-um ago. At that time the prevailing image was entirely influ-enced by Ptolemy's great *Geographia* (ad 90–168 c.). Ptolemy conceived the Earth as spherical; the problem he bequeathed

[1]W. H. Babcock, *Legendary Islands of the Atlantic* (New York, 1922).

was a lack of enough known world to cover the sphere. The whole of the globe was centred, as the name implied, on the Mediterranean. Every Greek sailor knew that once one had sailed beyond the last points of land the sea just went on and on. What particularly frightened the Greeks, and therefore the European mind which inherited their philosophical tradition, was the idea of *void*. The sea's void, that infinitely dangerous blank beyond known land, was as worrying metaphysically as it was physically. The Greeks' idealisation of a static universe full of fixed entities, faithfully reflected in their mathematics, underpinned all that could be thought. The sea was a positive insult to this metaphysics, a naked opposition to it. Not only was the ocean of unknown dimensions but it was moving, unstable, in certain circumstances even breaking out of its natural confines. How then could this fluid void be mapped? How did one map an ocean when it was featureless? How did one represent an *absence* of topography?

To appease this Aristotelian *horror vacui*, mapmakers before 1500 resorted to a variety of devices, including that of coralling the ocean safely within a complete ring of imaginary continents like the zetetics' ice barrier. Until roughly that date maps of the world were entirely notional. Ptolemy had been rediscovered in about 1200, but *Geographia* was only a text. The features of the globe were extrapolated from it entirely according to the cartographer's fancy. Religious imagery was prominent, some maps depicting three continents which corresponded to the three sons of Noah. (The insistence that the globe assume a doctrinal shape lasted at least into the Age of Enlightenment. The French cartographer Robert de Vangoudy published a map of America in 1769 showing the land as divided by Poseidon among his ten sons, much to the amusement of Voltaire.) Other shapes assumed by the world's land masses at mapmakers' whim were neat crosses, caskets, and the tabernacle. Until a certain date the function of land on mappemonds is to express the wish that the physical world should conform to theological or aesthetic categories. Underlying this, though, runs an anxious desire to frame a linked and anarchic series of voids into distinguishable oceans. When this is done the picture reverses out and instead becomes a map of land with bits of sea in between. It is a profound relief.

Mayda and all the dozens of other islands in the North Atlantic were the surviving, mobile fragments of conceptually necessary but imaginary land. This is precisely Claude Lévi-Strauss's idea of the 'floating signifier.' He argues that in culture there is always a need for certain concepts and expressions in order to soak up any excess of existence which has not yet been turned into words. It is the analogue of the algebraic concept of *nought*, which it is necessary to have before other things can be deployed.[1] The Island of Mayda's function was to be non-existent, to blot up an excess of vacancy, until something more solid turned up. Its poignancy is that even when it had been rendered redundant cartographers were loath to part with it. This was no doubt a matter of pride as much as sentimentality. Mayda was doomed to wander in an oceanic oubliette like a melting ice floe until being covered with the map's legend. When in some future edition the legend was moved, Mayda was found to have vanished.

Mythical or badly misplaced islands were not, of course, confined to the North Atlantic. The parts of the world remoter from Europe which were explored, mapped and named later had their own share which lingered correspondingly longer, even though by the mid-nineteenth century navigational methods were very reliable. Dougherty Island was believed for over a century to lie to the south of Australia. It was frequently reported at its given coordinates and in 1893 a New Zealander, Captain White, claimed to have sailed entirely around it, saying

[1] See Brian Rothman, *Signifying Nothing: The Semiotics of Zero* (1987)

there was no other land for 1,100 miles. After that it vanished, but its image endured on US charts until 1932. The same went for Podestà Island, named for the Italian ship which discovered it, the *Barone Podestà*. It was sighted in the South Pacific, some nine hundred miles off the Chilean coast and the ship's master wrote down its position. However, since his name was given as Captain Pinocchio it is perhaps not surprising it was never seen again, though it survived on some charts until 1936.

More inconvenient was the case of Sarah Ann Island, which ought to have formed part of the Gilbert Island group in today's Kiribati. In 1932 some American astronomers decided it would be the ideal spot from which to view a long-awaited total eclipse of the sun. The US Navy went off to have a look at the island and report back regarding the difficulties of accommodating scientists and their equipment on it. Despite searching, however, they failed to find Sarah Ann and the observations had instead to be made from Canton and Enderby Islands, five hundred miles to the east.

Of all such islands maybe the longest lived were—or are—a couple far down in the Pacific, south-west of Tierra del Fuego, Macy's Island and Swain's Island. These had vanished from practically every known chart by 1939, but they are still there in the 1974 edition of the *Soviet Atlas of the Pacific Ocean*. It is not clear which country they belong to; nor, indeed, to which era. One of them even persists on a John Bartholomew map in *The Times Atlas of the World* (the 1986 reprint of the seventh edition of 1985).

If this seems like a catalogue of earnest errors, the case of Hunter Island is one which from the start ought to have aroused suspicion but which was long treated very seriously in some quarters. In 1823 Captain Hunter of the brig *Donna Carmelita* claimed to have found an island some three hundred miles north-west of Fiji. Not only did he establish its exact position, he landed and found an island which was intensively cultivated and lived on by a tribe of 'highly developed' Polynesians. These natives had certain peculiarities. They all had their cheeks perforated in weird patterns and the little fingers of their left hands amputated at the second joint. For many years passing skippers kept their eyes skinned for Hunter Island, since apart from anything else the captain's account had intrigued a good few anthropologists who had never before heard of a Polynesian tribe with these curious characteristics. Alas, the island was never seen again. The idea that the entire thing might have been invented as a joke by a bored captain with a gift for subversive fantasy might have crossed somebody's mind on learning that his original—and private—name for the island was Onaneuse.[1]

'Tiwarik' and Onaneuse and, for that matter, all the world's chimerical islands have things in common which do not depend in the slightest on notions of 'objective reality.' To the gaze beneath which they once fell they had an absolute existence. St. Brandan's exercised its significance on the religious imagination of the time, while Seven Cities with its gold sand and Buss with its 'champaign country' stood as lands of promise whose only fault was to slip further out of reach as the centuries passed. Onaneuse, since all jokes are serious until their inventors begin to laugh, maybe stood for something which had been at the back of Captain Hunter's mind long before he ever went to sea. It might be oddly reassuring to invent a private erotic idyll, give it precise bearings and then watch other people search for it in vain. While the Captain laughed the scholars accorded it a status he would never have dared allow himself to claim. They did his work for him even as they failed to find it, and since he had known all along it wasn't there he had the last laugh as well.

[1] See Karl Baarslag, *Islands of Adventure* (London, 1941) and Henry Stommel, *Lost Islands* (1984).

No doubt islands draw some of their peculiar significance from the dozens of cosmogonies which begin with a watery chaos out of which land emerges. Any emergent land must initially take the form of an island, so the island stands as the archetype of land. As to what this proto-land might contain would depend on when it was first spied, and by whom. Paradise, treasure and naturalists' nightmares were variously seen as appropriate, but the nearer our own century was approached the more an explorer, an adventurer or a philosopher might expect the proto-land (glimpsed tantalisingly in the parting of a fog bank or glittering in the objective of his telescope) to contain a domestic order reassuringly old-fashioned as well as exotically unlike any known society. Things are different nowadays. Nobody any longer expects to find a place where the people are nobler, sexier or just better behaved. Wistfulness has been replaced by a certain hard-nosed quality. If you can't find them, you found them. Were there such a thing as an endangered species of land, the island would be it. Far from being proto-land it is coming to feel like a last land. The whole concept of the island, which until recently was implicit with all manner of promise, is now redolent of loss.

'Tiwarik' will go on existing only as long as its author. Unlike the island to which I attached the name, it is not contingent on Japanese developers. Somewhere its grasses still blow in the wind. Six or so years ago, when it was its old self, I ended my description by calling it an act of the imagination. This will always be true of places which at last become properly real to us.

> It is not its grasses my feet have trodden nor its little coastline I have so lovingly followed, and neither does it retain any trace of me. There is another island locally known as 'Tiwarik' but it is only an exact facsimile, a fly-spit on the map of the objective planet which we agree to inhabit.[1]

[1] Hamilton-Paterson, *Playing with Water* (1987).

II

Objects of Desire

Four things about small islands:
1. They look like objects and, hence, like property.
2. The concept of the 'private island' satisfies most people's major fantasies.
3. The effect of islands is almost wholly regressive.
4. An island's boundaries can never be fixed.

1. They look like objects and, hence, like property

Everyone looks at an island, whether consciously or not, much as a tyrant eyes a territory. It takes a long time to have any relationship with a land or a country, but the mere sight of an island from an aircraft's window or a ferry's deck mobilises the beginnings of possessiveness. The place is small enough to treat with, to become familiar, to exhale an air of exclusivity, even if it is quite nondescript. A slight grammatical shift can mark either social desirability or small size—usually going together. Thus, one has a house *in* Malta, but a bungalow *on* Gozo. He lives in Jersey, she on Sark. (But they have a house on Long Island as well as one in Jamaica.)

This unit of land which fits within the retina of the approaching eye is a token of desire. The history of the Isle of Buss shows this desire working so strongly that successive mariners appropriated a portion of a long coastline and changed it into the island they would have preferred to discover. To have happened upon an unclaimed continent while lost in a small fishing smack would have been inconvenient, but to have found an unknown island was both manageable and enviable. How, then, could its discoverers have extrapolated a self-contained shape from a length of coastline? How were they able to draw the fictitious 'back' of this 'island' which remained forever as hidden and theoretical as the dark side of the Moon? Mediaeval cartographers often solved this problem by giving the Atlantic islands stylised shapes: circles, clover leaves, rectangles and crescents. The Isle of Mayda retained its crescent or indented circle shape on map after map, and eyewitness accounts of it seemed to conform to this outline with remarkable faithfulness. Quite possibly this reflected its rumoured Islamic origin.

There for the taking . . . Ever mobile, for several hundred years the lost islands of the Atlantic might bob up anywhere from behind freezing mist, in a hurricane, or during a search for somewhere else entirely. The point was they could be possessed at the drop of an anchor, named for a vessel, claimed for a monarch. Even today, visitors and holidaymakers may 'discover' an island which becomes 'theirs' in respect to their friends, envious neighbours, peers.

Icebergs, floes and ice islands also form a particular class of islands in that they are both mobile and temporary. They look like and are objects, and are sometimes colonised by Eskimo hunters and teams of scientists for varying lengths of time. Several islands made of shelf ice and far larger than the Isles of Wight or Man have provided stable bases for research stations for ten or more years at a stretch. The question of possession is another matter, though. If they are inside territorial waters there is no problem; but many icebergs carry with them a vast tonnage of boulders and other morainal material and one might wonder to what extent they go on being part of the nation in whose territory they were calved. Canadian soil and Canadian water presumably made up the iceberg which sank the *Titanic*; but while the Canadians would have retained full rights over it while it was in their waters, would they automatically have ceded all responsibility once it had left? Presumably so, otherwise the White Star line could possibly have brought an action against the dominion for negligence in allowing pieces of its sovereign territory to go drifting away out of its control.

The boulders carried by such icebergs and released as they melt often end up thousands of miles away from their place of origin and in the early deep-dredging oceanographic expeditions of the mid-nineteenth century caused geological confusion. Were such boulders now discovered to contain valuable rare minerals in exploitable quantities, perhaps Canada and Denmark might pursue a legal claim granting them exclusive mining rights over their pieces of rogue territory shed from Newfoundland and Greenland, a claim which would also exonerate them from blame when anyone accidentally rammed one of their melting assets.

2. The concept of the 'private island' satisfies most people's major fantasies

This assertion may be true only for our culture; but as Western culture in general seems regrettably set eventually to subsume most others it is probably at worst a truism. The 'private island' fantasy is simply one expression of the urge to define, annex and defend territory. It is clear that in this context 'island' can as easily mean any patch of land anywhere, even a mere house. This is especially noticeable in England, where his home is only half-jokingly referred to as 'the Englishman's castle.' The apotheosis of this is a place like Loch Leven Castle in Kinross, where Mary Queen of Scots was imprisoned and from which she escaped in a rowing boat.

This consists of a castle on an island in the middle of a lake which is itself on an island in the north-east Atlantic. The idea embodied in this arrangement can be expressed graphically by a series of concentric rings: circular boundaries nesting within one another, lines of exclusivity and defence which intensify in power the more they approach the centre. The average mediaeval castle, in default of a handy lake with an island, had to make do with a moat, thereby becoming an artificial island.

There is certainly a tendency, perhaps more pronounced in some cultures than in others, to make 'islands' on dry land. In Tuscany, for example, the natives increasingly resent the habit of foreigners (meaning both non-Italians and non-Tuscans) of buying up pieces of their countryside, fencing them off and forbidding all access to them by locals. As with the villagers of 'Sabay,' the resentment is mainly twofold. The Tuscans do not like their immemorial rights to hunt, gather, stroll or otherwise come and go suddenly abrogated, nor what they have always considered part of their horizon to be out of bounds to them. At the same time, they are put out that it is not they themselves who had the financial liquidity to take advantage of the boom in local land values. Had they done so it is open to anyone to wonder whether they might not with alacrity have assumed the grandiose mantle of landowner and as swiftly put up fences and given large, mean dogs the run of their property.

The 'private island' remains the correlative of a particular dream. Islands are at once objects of desire and a locus for desires. The dream embodies fantasies of autonomy, independence, security, sex, grandeur, individuality and survival, in recognition that modern metropolitan and suburban life connotes powerlessness, dependence, defencelessness, frustration, lack of status, anonymity and a general feeling of expendability. In waiting rooms, people eye colour advertisements in *Country Life*, aerial views of yet another Scottish island about to come under the auctioneer's hammer, while an easily decoded dream crosses their mental retinas and glazes their eyes. '*Estimated price: £750,000.*' The same dream leaks into all sorts of stories and films set on private islands where the unities of time and place can be rigidly controlled. These may be tales of man-hunts with the narrator-guest as the next quarry; reigns of terror; ghoulish experiments; masterminds plotting the world's overthrow from their flamboyant yet top-secret lairs; elaborate erotic baroqueries. Science fiction carries the dream on, being full of expansive futures in which the rich and powerful own private planets, while even the moderately wealthy may aspire to a humble asteroid as the site of a kingdom, retreat, hideout or love nest.

Nor is the dream confined to adults. In their coastlines, as in their potentiality, all lost islands go on reappearing in the maps which every powerless schoolchild draws.

3. The effect of islands is almost wholly regressive

Islands infantilise people even as people idealise islands. Those with appetites and no souls think they would be safe from the eyes of the world. Those with soul and little appetite believe they can fall under an island's benign and teaching gaze.

The island repeats a fantasy of human beginnings. The foetus—castle of the ego and keep of the soul—is effectively an island for the first nine months of life, entirely surrounded by an amniotic moat and connected to the mainland only by an umbilicus. Soon afterwards the playpen becomes an island, probably the most fabulous of all. Not only does the infant command its every square foot, he commands the world which his own supreme frontiers deign into being by marking off. His shores, his limen; and so by extension his ocean, his continents, his world.[1] Moreover, the fantasy of a private island always takes on that infantile characteristic of absolute flexibility in being able simultaneously to stand for almost any desire and to serve as the ideal locus for practically any fantasy. For islands are also sexual places because they have the air of being extra-legal, extra-territorial, out of sight and censure. Every so often a film appears depicting torrid intimacies among the conveniently marooned. For this cinematic purpose the island must be tropical and the state of undress constant. It would not be at all the same for two nubile castaways to find themselves stranded in the Bering Sea.

The island is thus the perfect territorial expression of the ego. As such, it is all too easily a metaphor for the individual. Sometimes the metaphor is used at one remove, so the island takes the place of a wise alter ego. The message here is that man learns by true experience of himself. The lessons may be practical and moral (as in *The Swiss Family Robinson* or the story of Alexander Selkirk) or spiritual (as in Richard Nelson's *The Island Within*).

The infinitely flexible nature of islands, of their being at once safe and adventurous, constraining and boundless, erotic and polemical, has made them ideal destinations in a long literary tradition of imaginary voyages. More than 1,000 years before Homer there was a twelfth-dynasty Egyptian story about a castaway on a marvellous island, and Plato's account of Atlantis functions as a kind of blueprint on which he might later have constructed a more complete utopia. When Sir Thomas More produced his own original *Utopia* in 1516 he

---

[1] Part of the island's haunting quality may be because its exclusivity reminds us of the family as we once saw it through infant eyes: self-contained and self-sufficient. A family's underlying sadness resides in its conspiracy of immortality. When decades later we come to look at it with an almost-stranger's eyes, a family relic such as an old tablecloth now stands poignantly revealed in its faded colours and moth holes as having always been both altar cloth and shroud.

put fresh life into an ancient genre. The dignity of his Latin must have induced many a lesser writer to indulge his own intellectual fantasies under the disguise of gravity, for the literature of the next three centuries abounds with all kinds of utopias and ideal commonwealths, most of them sited on imaginary islands. (At this point, and quite gratuitously, I wish to note an allegation that Sir Thomas More 'used to thrash his grown-up daughters with a rod made from *peacock-feathers*'.[1] Without bothering to try and put a finger on it more precisely, one feels this sort of behaviour is not inconsistent with thought about islands and ideal societies.)

It is curious there was no discussion in English of the imaginary voyage as a genre before the nineteenth century. Indeed, there was not even any recognition that it was a literary type worth discussing. In France, on the other hand, there were all sorts of studies and by 1787, when Garnier's remarkable *Voyages imaginaires, songes, visions . . .* was published, he was able to subdivide his classification of Allegory into a whole variety of islands, among them an *île d'amour*, an *île de la félicité*, an *île taciturne*, an *île enjouée*, an *île imaginaire* and an *île de portraiture*. After Crusoe's great success in France, several imitative *Robinsonades* showed what man might be capable of when thrown entirely on to his own resources, whereas adventures on an *île inconnue* tended to depict what happened to a domestic society cut off from the rest of the world. In this respect they constituted something of a counterpart interest to that in feral children (such as Victor, the wild boy of Aveyron), around which at that time all sorts of arguments revolved concerning what exactly constituted 'the natural' and 'the civilised'.

The genre still exists. The French writer Georges Perec in a novel published in 1975[2] uses an imaginary island off Cape

---

[1] Magnus Hirschfeld, ed. Norman Haire, *Sexual Anomalies and Perversions* (London, 1959), p. 396.

[2] Georges Perec, *W ou le souvenir d'enfance* (Paris, 1975).

Horn as the setting for a fascistic society obsessed with sports. And what else is one kind of science fiction but a convenient locating of utopias and dystopias off-Earth on imaginary planets which are, from our perspective and by any other name, islands in space?

The effect of islands is almost wholly regressive . . . This is most true of prison islands since there is nowhere more regressive than a prison. 'Regressive' should not, of course, be read as a synonym for 'comforting,' although many brutalised and institutionalised people find incarceration reassuring. The objections are obvious. Did Napoleon 'regress' on St. Helena? What was so comforting about *le bagne* which induced 'Papillon' to escape Devil's Island? Was the camp on Blood Island in some way deeply cheering? But 'regressive' applies as much to the behaviour of a prison's governor and guards as to that of its inmates. It refers to the effect on everybody concerned which all institutions exercise and penal institutions in particular. Punishment is by its nature regressive, and prisons usually involve extremes of pettiness, brutality and sexual licence: the normal ingredients of infantile behaviour. Seen in this light, Devil's Island was something like a nineteenth-century English public school with mosquitoes and a guillotine. A succession of the prison's governors encouraged their charges to settle into some variety of homosexual marriage, jointly tilling a small garden and sharing its produce. This was considered an effective antidote to the yearning for escape, a progressive piece of penology known locally as 'the cucumber solution' (*la résolution du concombre*).[1]

At the very centre of imprisonment's concentric rings is solitary confinement: isolation (from *insula*), which if inflicted for long enough on the wrong individual may cause a regression from which it can be hard to emerge. It all depends on

[1] See Chapter 7.

one's position. From the authorities' point of view isolation is the place where the community expels the individual. From certain individuals' points of view, solitude is what they long for to escape the community. Their yearning may even be for the numinous, invisible spot in the centre, identity's apotheosis and vanishing point.

4. An island's boundaries can never be fixed

Chapter 1 hints at some of the problems international law has in trying to fix Exclusive Economic Zones. In terms of absolute position, boundaries, like the Earth's crust itself, can remain fluid. Thus when I hypothesised the case of an island's EEZ happening to cross the edge of a tectonic plate and there-fore either widening or shrinking, the response at IOS was a scathing 'Good God, they can't measure to the nearest nauti-cal mile, let alone to the nearest centimetre.' Over time, though, these centimetres add up. In 1492 when Columbus crossed the Atlantic, America was twenty metres nearer to Europe than it is today.

One may multiply almost indefinitely the special cases of volcanically inconstant islands, islands whose coastlines are hard to define, artificial islands and self-proclaimed independ-ent principalities sited on World War II anti-aircraft towers six miles off the Essex coast (a perfect example of an island as egoic headquarters).[1] There are islands like Rockall which seem to have no purpose but to act as a focus for legal wran-glings, their sole value being as pegs in the ocean around which a lucky winner might draw an EEZ. Japan has recently been driven to considerable feats of technology in order to rescue its

---

[1] 'Sealand,' otherwise known as Roughs Tower, turned by Major and Mrs. Roy Bates into a 'state' with its own constitution, flag, stamps, coinage and passport. See *The Independent*, 24 February 1990.

own eroding islet of Okinotorishima at the extreme southern tip of the Japanese archipelago. At high tide Okinotorishima consists of two lumps of coral respectively three metres and five metres wide. Without these little rocks Japan's territory would end at Iwo Jima and she would lose 154,440 square miles of EEZ and with it all exclusive fishing and mineral rights. With each typhoon the rocks erode a bit more, so the Japanese have spent upwards of $300 million to enclose the island in a concrete and steel protective wall. It is important that no part of the wall should touch the corals because this would turn Okinotorishima into an artificial island and disqualify it from having an EEZ. As a professor of international law, Soji Yamamoto, says: 'Territory is not something that exists naturally. It must be obtained by a country's efforts.'[1]

Meanwhile, if sea levels rise as predicted over the next half-century, what of island boundaries then? Will the original outlines of a largely submerged Maldives be maintained on maps as much out of respect as for legal or navigational reasons, so sightseers can peer down from the rail of a boat at a lost land as they might today at a sunken battleship?

Other kinds of island boundary are blurred, too, especially when there seems to be no logical or geological reason for assigning a particular nationality. Corsica is historically, linguistically and geographically more Italian than French. The Channel Islands might quite as easily be French as British and, with their degree of autonomy, form part of that odd, anomalous category of offshore tax havens which are neither fully integrated nor fully independent. The Isle of Man also falls into this category. Such islands generally have a certain diehard quality and are as much leftovers of an older social order as they are of a former geological configuration.

It is archipelagos and chains of islands which are so often

[1] See *Nature* 333, (9 June 1988) p. 487 and *Fortune* 122 (24 September 1990) p. 12.

the geographical versions of displaced persons, holding at best a temporary passport. The Sulu Archipelago is a perfect example. One has only to sit on the wharf in Jolo to be prey to a literary sense of unreality. The waterfront of huts built on stilts over the sea, the lumps of islands at every distance, the decaying ferries and wooden launches full of fish and copra and red logs: allowing for a lack—but not complete absence—of sail, it is Conrad's horizon still, and filled as ever with the dreamy tropic energy which slops across all boundaries. Politically, Sulu is part of the Philippines right down to Tawitawi, at its closest reaching to within a few miles of Sabah in northern Borneo. However, in 1981 it became part of an autonomous region with special barter trade rights between it and Malaysia. This was in response to the long and bloody war waged by the Moro National Liberation Front (MNLF) against what they thought of as an attempt by Christian Manila to oust or dilute Islamic culture and greedily expropriate whatever it is that governments habitually do greedily expropriate.

On Jolo jetty Conrad's azure map lies ahead, while immediately behind is a troubled, dark green backdrop of conspiracy and heavy weapons. Conrad would have recognised that, too, since men here have always gone over-armed. A drunken fight can lead within minutes to mortar rounds. The Republic of the Philippines, with its implied promise of centralised law and institutionalised order, covers Sulu with a cartographer's fiction. The islands of the archipelago are defined and individuated by language, usage, tribal politics, gangs, bandits, even pride. They are criss-crossed by the interests of disparate ethnic groups, trading links, smuggling, piracy, local tyrannies, fishing, seditious movements and intersecting anarchies. In such places official boundaries vanish entirely unless drawn fleetingly by the wakes of navy patrol craft or coastguard cutters. I once went on a week's fishing trip in an open boat from Palawan southwards. We fished for lobster off Bugsuk and

Balabac islands, sleeping at irregular intervals wedged into the bows or on occasional dry land. I lost all sense of time and position. On the way back I discovered we must have spent one night on Borneo. The same thorns, mangroves and littoral clutter, it had seemed nowhere different.

The example of 'Tiwarik' and its grandiose conversion from nondescript islet into businessman's fantasy, vaulting a strait to appropriate a chunk of mainland, is a reminder that if physical and legal boundaries can often only be fixed with great difficulty, then areas of wish can never be clearly demarcated. The greed of dreams is to expand into any space denied them. Fantasies, daydreams and dreams flap about as they may, but they all roost in the unconscious and share its logic, which is that there are no contradictions. Just as sexual fantasies can simultaneously involve a single person and many, at once watching and being watched, doing and being done to, so may an island be experienced as both small and infinitely large, part of the land and part of the sea, sheltering and exposing, terra firma and freakishly unstable. As they melt, icebergs may become very unbalanced, often rolling completely over without warning. The private nickname 'Tiwarik' was prescient, for the word means 'upside down' and something about the place had always suggested rich possibilities of inversion. Never had I thought to see it so thoroughly stood on its head.

All this points to a boundary which as yet appears on no map but which is ever more real and ever more cavalier. If one considers ex-'Tiwarik' as a piece of terrain which has recently been annexed for the convenience of a foreign 'lifestyle,' then probably there can be no such thing as a holiday island. Once an island becomes a resort it ceases in some essential way to be an island and turns into an extension of a mainland, even if that is half the world away. Nor need this mainland even exist as a sovereign country. It is enough to comprise that fictional

international place to whose citizenship so many lay claim or
aspire: that stateless state of BMWs, Chivas Regal and Dunhill
lighters; of treasure and pleasure and leisure, guaranteed
moth- and rustproof. Since much of this fabled place lies in
chilly latitudes, it needs to push out long, tentacle-like penin-
sulas into warmer climes: Vacationland, embracing all sorts of
otherwise under-used bits of lesser countries in the sun. It simply kits these out with its own standard furniture (scuba gear,
water skis, hang gliders, beach barbecues, rock music, drunks,
whores). And suddenly there are no more islands, only scattered slabs of a single moneyed empire joined each to the other
by something solider than water. Or so it seemed, waiting for
'clearance' in the new beach pavilion on what was once
'Tiwarik.' This building is now a border post. Under
Philippine law no private individual may own any part of the
foreshore below a point thirty feet above the mean high-tide
line. It is national property. Hence anyone may land on the
Fantasy Elephant Club's beach. But where once by civilised
common usage they were free to wander the rest of the island,
they now have to be vetted by an immigration official in a blue
uniform and wearing a gun. The beach has changed from a
haven into a frontier.

There is one last kind of island, one whose elusive presence
flickers at the edge of vision, quick as fish. This is the imaginary island faithfully mapped in every psyche, mostly unsuspected, infrequently discovered, even more rarely inhabited.
An outcropping of the self, it lies across a treacherous strait
which discourages acquisitiveness and, even on clear blue days,
may have vanished as if it were roaming the oceans in search of
the one worthy inhabitant. Then on a rare day the rare person
wakes and it has swum out of the corner of his eye and stands
before him. On such a morning it takes no effort to cross over,
paddle flashing in the sun, until the skiff's bows nudge grind-
ingly into the shore.

And then what pleasure to set up a hut, a fish drier; to pare things back to water and light, to knives and spearpoints, to order and silence! *All men have an island*, Donne should have said, for a suspended wheel rim being beaten in a cement block chapel on the distant mainland ought to tell us no more than the fish curling and flapping between our hands, bleeding rusty threads into the sea. That steely tolling from across the water brings no news, nothing we do not already know as later we climb the headland to watch soft dusk well up over the world's rim and efface the ocean below. It is not interesting to tot up the sunsets seen and perhaps to come. Those deaths, our deaths, are not plangent affairs but matters of geology. We are all at best marginalia in another era's fossil record. Go down to the hut instead through a drift of fireflies. Light the lamp, cook rice. There is nobody else on this island; there never was and never could be. Outside, the waves wring green flashes from plankton. The great mineral machine turns its fluid gears. The firefly in the thatch tugs us into its gravitational field.

III

MARGINALIA:
The Isles of Britain

Did Britain cease to be an island at some debatable moment between the end of World War II and the construction of the Channel Tunnel? Until 1939 Britain had lavishly exported its boundaries in terms of Empire, and while much was made of the idea of this 'tiny island nation' ruling a good part of the globe, it did still retain certain characteristics of an island. Among these were the world's largest maritime presence, a thriving fishing industry around its entire coastline and an attitude towards the rest of the world which can only be described as insular. (The putative headline *Thick Fog—Continent Isolated* puts it quite succinctly.)

The war provoked insular rhetoric, ranging from endless Shakespearean tags about 'this precious stone set in the silver sea' to re-emphasising the tininess of 'this island nation,' although not now in comparison with the world it ruled but with the Hitlerian might threatening to crush it as it 'stood alone in the dark days.' The white cliffs of Dover stopped being rather striking beds of calcareous plankton and took on the quality of Britain's boundary, ramparts, bulwark, palisades. A few miles of cliffs came to stand for an entire spiritual seaboard which might not be violated. Had that portion of Britain's coastline most adjacent to the rest of the world been as low and unremarkable as it is in parts of Essex, Suffolk or Lincolnshire, the rhetoricians and songwriters would have needed to come up with a different trope altogether.

After the war, Britain with its depleted spirit and collapsed economy became vulnerable to the very things which erode islands: links with foreigners and—as the economy improved—increasing travel abroad by the islanders themselves. Somewhere between 1945 and 1990 Britain lost consciousness of itself as an island. The sea which surrounds it (and which only poetic fancy could ever have described as 'silver') now plays virtually no role in its thinking or its economy. Its naval and merchant fleets are shadows; its fishing industry is ravaged; even its own people find the beaches of Spain and the Greek islands more congenial than those at home. So when the two pilot tunnels, one from England and one from France, met beneath the Channel in 1990 and Britain became technically joined to the rest of Europe by dry land, the event seemed to provoke remarkably little upheaval in the nation's psyche. King Arthur, Drake and Churchill slumbered on, unmoved. The country was already part of the EC, with the free movement of member citizens guaranteed. Insular attitudes still abounded, but something had changed. A generation had grown up which knew and cared nothing of precious stones set in silver seas and would in any case think the description fitted Barbados or Mykonos far better.

It is probable that only a war which threatened Britain directly could ever resurrect the insular rhetoric. In his speech in *Richard II* John of Gaunt did mention infection, however. ('This fortress built by Nature for herself / Against infection and the hand of war . . .'). It so happens that Britain does still jealously guard the ghost of its island self by means of a last and potent symbol of foreign contamination: the fear of rabies. Ironically, it is mad dogs which give the Englishman his final solitary status.[1]

---

[1] Since the UK government's Pet Travel Scheme (PETS) was introduced in 2004, pet dogs and cats may enter Britain from qualifying countries without being quarantined provided they have been microchipped, vaccinated against rabies and have waited at least six months before a blood test reveals a suitably high level of antibodies. Although the possibility of mad dogs still underwrites his fear that contamination could arrive at any time from continental Europe, today's Englishman is apparently more alarmed at the thought of unquarantined human immigration. Hence the UK's opting-out of the Schengen Agreement.

There is about the swimmer a sentimentality, or self-pity, which disgusts him even as he finds himself thinking that surely he ought to have earned a reprieve. All that close attentiveness to the sea over the years, to this ever-yielding, stony-hearted medium which has him in its embrace—it surely cannot have been wasted. There must be something he has learned from it, some subliminal message from his ancestral home, instructions for survival? The idea is fatuous but persistent.

He is beginning to tire. Not of staying afloat, since this is effortless, but of trudging the water to stand higher, of spinning to keep every horizon constantly in view. More and more he allows his face to hang in the water. Through the glass panel of his mask his vision lengthens past the rope's end twenty feet below his ankle. He no longer sees the prismatic chips of phytoplankton, the blazing motes and jellies as they drift past his face. Now he believes he glimpses shapes far beneath, not predators but bulks of deeper purple as though . . . Why not? This archipelago is full of hidden reefs, its contorted seabed thrusts up unexpected pinnacles to within feet of the surface in the middle of nowhere. There could easily be a coralline peak, a ledge, even a plateau over in that direction away from the sun where the water does seem to deepen its colour as if a little further down there lay a solidity . . .

Away from the sun? The swimmer jerks his head up, gasping and squinting painfully at the blazing disc overhead. Is it not past its zenith now? Has it not begun to sink? May this illusion of a darker bulk be nothing but his own shadow, cast as he has so often seen it in late afternoon? No; ridiculous. It is not late afternoon, merely maybe a few minutes past midday. He looks downwards again and in a while his eyes adjust from the dazzle and once more he thinks he can pick out an area of deeper tone. So convinced is he that he begins swimming towards it, slowly, so as not to give the impression of having finally picked a direction or of expecting very much. Now and again he glances around to tell his invisible boat where he is going. It seems to him that all will be well if it turns out to be a lonely reef he is heading for. However small, it will convert the ocean at that point into a shallow sea. He would then, as if by magic, be in his depth in 20 feet of water. Or at least, hovering as if in air above the unknown but familiar city, he would be close enough to feel its dwellers might intercede for him, present his case for survival at some court of marine jurisprudence . . . As an idea it is better than nothing; even a glimpse of a reef would sustain him. Besides, reefs were mysterious and deceptive places whose greater being remained hidden. If one kept one's eyes on a reef under water and followed its avenues it had a habit of turning into a shore. The swimmer had experienced it a thousand times. Might not this one, despite an apparently yawning horizon, somehow work the same friendly trick?

3

Reefs and Seeing

Figures chosen by Darwin to illustrate his theory of coral-reef formation:
(i) A small island (Bolabola) surrounded by a barrier reef
(ii) A 'lagoon island' or atoll whose central peak has sunk, in this case (Whitsunday Island) leaving a rare example of a complete ring of land
(iii) Goniopora columna
(iv) Porites mordax.

I

C oral was used variously as jewellery, medicine (ground up like mummy-powder) and inlays in cabinet-making. The precious red variety was traded for thousands of years without anyone being able to say precisely what it was. It was found in the sea, certainly, but fishermen and dealers were studiedly vague in order to protect their source. It presented a problem in the eighteenth century, however, since with the development of natural history it finally became necessary to say whether it was animal, vegetable or mineral.

Perversely, it exhibited characteristics of all three. At a time of industrious taxonomy this was unsatisfactory and exasperating. Coral reefs were notorious from the lyrical and often ambivalent accounts of sailors, who could one day view them as luxuriant herbaceous borders of surpassing delicacy and beauty, and the next as the most treacherous threat the tropical oceans could offer, ready at the first sign of a dozing lookout to tear the bottom out of a ship. The imagery used to describe reefs reflected this ambivalence; it also reflected confusion about the substance being dealt with. Horticultural images were common, and R. M. Ballantyne's view of a reef in his novel *The Coral Island* (1858) as a sublime rock garden was by then almost a cliché, one of the two commonest ways of seeing coral. The other, somewhat later, was to view it as architecture. A reef was a fantastical cityscape, habitation for King Neptune and his courtiers (that is, when they were not housed in an Atlantean version of this kingdom, whose style tended towards Hollywood Athenian). Perhaps certain illustrations of science-fiction cities, in their lack of any unifying architectural style and their proliferation of bizarre organic shapes, derive ultimately from the iconography of reefs.

Until about 1750 zoophytes—that is, plant-like organisms such as anemones—were classified as vegetable. The very terminology (animal-plant) reflected uncertainty. At that time 'the animal' was distinguished from the other class of living creatures as having the capacity to move at will from place to place. It was a long while before the agents responsible for coral were identified as zoophytes; but when they were, many saw them as vegetable since, like marine plants, they were attached to the substrate by disc or tubular fibres. What was more, the branching arms of coral looked convincingly arborescent.

This classification was challenged by mineralogists. They argued that many zoophytes were hard and stony, a description which fitted no known plant. Clearly they were rocks, made of calcareous and clayey sediments 'moulded into the figures of trees and mosses by the motion of the waves, by crystallisation, by the incrustation of real fuci [seaweeds] or by some imagined vegetative power in brute matter'.[1] It was not for nothing that many reef-building corals became known generically as 'madrepores' or mother of stone.

The animal theory had its supporters, too. According to one of its proponents, Dr. Vitaliano Donati, the 'stone' of coral was not stone at all, but *bone*. 'I am now of the opinion, that coral is nothing else than a real animal, which has a very great number of heads. I consider the polypes of coral as the heads of the animal. This animal has a bone ramified in the shape of a shrub. This bone is covered with a kind of flesh, which is the flesh of the animal.'[2] It is an ingenious and eccen-

[1] George Johnston, *A History of the British Zoophytes*, 2nd edn (London, 1847).
[2] V. Donati, *Phil. Trans.* (1757) abridg., xi, p. 83.

tric idea. An animal in the shape of a shrub is—if nothing else—the exact opposite of topiary, and suitable for a zoological garden.

When reviewing these conflicting theories one might imagine the issue could have been settled once and for all by any naturalist spending a few hours with a good microscope. Yet a certain Henry Baker, in his book *Employment for the Microscope* (1753), could only see evidence for corals as mineral. According to him they were accretions of salts leached out of nearby rocks by seawater and bonding themselves chemically to 'the calcareous and saline particles' of the water. At the same moment a London merchant named John Ellis began to use his microscope and came to different conclusions. Ellis was typical of a very English breed, the amateur naturalist, which was to flourish so valuably in the nineteenth century. Men with under-used intelligences and hours to fill would turn their attention to the fledgling sciences where there was still an immense amount of basic taxonomic work to be done, work which any dedicated amateur in quest of knowledge could perfectly well do. Country parsons were particularly given to this sort of hobby and alone accomplished field work and classification in a dozen disciplines including meteorology, geology, entomology, botany and ornithology. In addition to attending lectures at the Royal Society, Ellis, too, had a hobby. This was making miniature landscapes on sheets of paper with a collage of delicate fragments of seaweed and corallines. The more he used coral the more he wondered what it really was, finally putting it under his microscope. In 1752 he addressed the Royal Society, announcing 'that these apparent plants were ramified animals, in their proper skins or cases, not locomotive, but fixed to shells of oysters, mussels, &c., and to Fucus's.'[1]

[1] John Ellis, *Essay on the Corallines of Great Britain* (1755), Introduction.

Carl Linnaeus, the celebrated Swedish naturalist and father of modern taxonomy, was impressed by Ellis's work but nevertheless decided that zoophytes occupied a classificatory niche of their own 'between vegetables and animals: vegetables with respect to their stems, and animals with respect to their florescence.' Ellis would have none of this and wrote to the great man that he could not reconcile himself to 'vegetating animals.' He had a powerful sense of how things might partake of the characteristics of other things without fundamentally changing their nature. 'Artful people,' he wrote to Linnaeus, 'may puzzle the vulgar, and tell us that the more hairy a man is, and the longer his nails grow, he is more of a vegetable than a man who shaves his hair or cuts his nails; that frogs bud like trees, when they are tadpoles; and caterpillars blossom into butterflies . . .'[1]

In the mid-eighteenth century the Royal Society in London was to some extent the international arbiter of such things, and Ellis's view won more and more endorsement as correct. Many of the naturalists of the period had tried burning corals and found they obtained much the same smells and residues as from burning animal products such as skin, hair and bones. This seemed conclusive. After Ellis there was broad agreement that zoophytes, including corals and sponges, were indeed animal though there were some who, with Linnaeus, hedged their bets by preferring to think of them as 'intermediate beings partaking of a twofold nature'.[2]

This version of the dispute over corals turns out to be a very Anglophile account, implying as it does that it was all a matter of a handful of Britons working more or less in isolation. Such is not the case. Recent research makes clear it was a subject much discussed elsewhere in Europe a good half-century ear-

[1] Linnaeus, *Correspondence* (1821).
[2] Johnston, *A History of the British Zoophytes* (1847).

lier, especially in France and Italy.[1] It reveals that it was actually a Frenchman, J.-A. Peyssonnel, who first identified the living part of corals as animal. In 1706 the Italian Luigi-Fernando Marsigli arrived in the French town of Montpellier, where he found 'a thriving scientific community, serving the ancient university with its famous school of medicine and botanic gardens, a school of hydrography, and a Jesuit college'.[2] Having collected corals, he performed all kinds of experiments on them. He would take branches home and put them in seawater of the same temperature as at the depth where they had been taken. The next morning he found them covered in little white flowers. He took the branches out of the water and the flowers vanished again. He put them back and they reappeared. Over the next few months Marsigli experimented with adding different reagents to the water to see what effect they had on the flowers. Excited by this discovery, he evidently cannot have known that the concept of the 'flowers' of coral was already familiar and had actually been illustrated in a medical treatise five years earlier. There, the author had thought coral was a marine stone which somehow grew.

In 1724 the son of a French friend of Marsigli's became interested in these coral flowers and called on him. Jean-André Peyssonnel repeated all the older man's experiments together with some of his own. His findings convinced him that what everybody thought of as a flower was actually *un insecte* like a little octopus. He wrote a letter to the Académie des Sciences which was never published but whose ideas were later addressed by Réaumur and demolished in the lofty fashion of a grandee confronting the work of a junior. Réaumur declared that 'what Peyssonnel had taken for animals living in a habitat

[1] See Anita McConnell, 'The Flowers of Coral—Some Unpublished Conflicts from Montpellier and Paris during the Early 18th Century,' *Hist. Phil. Life Sci.*, 12 (1990), pp. 51–66.

[2] McConnell, 'The Flowers of Coral'.

of their own creation, analogous to the wax combs of bees, were merely insects infesting the coral stems . . . Réaumur himself believed coral to have a dual nature—a vegetable "bark" growing on a stony core."[1] Snubbed, Peyssonnel waited until 1751 when he sent a copy of his *Traité du Corail* to the Royal Society in London. This was translated and extracts published in 1752, the year John Ellis addressed the Society with his own theory of corals being 'ramified animals.' Peyssonnel had anticipated him by thirty years.

The idea that corals were not colonial invertebrates but a strange hybrid species lingered for a long time, and even the authoritative George Johnston a century later dissented with this view of sponges because they lacked polyps. He asserted they were plants and quoted in his support a certain Professor Owen writing in the *Lancet*: 'If a line could be drawn between the animal and vegetable kingdoms, the sponges should be placed upon the vegetable side of that line. Locomotion could be no proof of animality; for it was well known that the sporules of some cryptogamic plants possessed very perfectly the power of motion.'[2] Even in the twentieth century some people still found the living—let alone animal—nature of corals hard to accept. Vice-Admiral Boyle Somerville records an instance when crew members of HMS *Penguin*, ordered to take soundings in the Kermadec Deep near Tonga, became obsessed with the beauty of some reef formations and went to great lengths to snap off and bring back several large chunks as souvenirs. These were winched aboard the ship under the indulgent eye of the skipper (who had actually been in the old *Challenger* in the 1870s) and young Lt. Somerville. There is no indication that either of these men knew any better, and they seemed as surprised and disillusioned as their crewmen when

[1] McConnell, 'The Flowers of Coral.'
[2] *Lancet*, 871 p. 225.

after twenty-four hours in the ship's hold the corals sent a stench of putrefaction throughout the vessel.[1] No doubt even today tourists by the thousand make the same error.

Susan Schlee suggests disarmingly that part of the reason why so many nineteenth-century oceanographers were fascinated by corals was that most of them came from cold, wet, northern climates and coral reefs are largely found in balmy, tropical and subtropical latitudes.[2] Once it was agreed that corals were animals, there was a new problem to be addressed, itself a promising field of research. This was, how could atolls have grown from the deep seabed when the reef-building polyps could not survive below about two hundred feet? There was a suggestion that maybe they built on the lip around the craters of undersea volcanoes, but that left geology with the task of explaining how there could be so many volcanoes of nearly identical height, all of them conveniently reaching to within one hundred feet of the sea's surface.

There are three main kinds of coral reef: barrier, fringing and atoll. A barrier reef generally follows the coastline, often at a considerable distance, converting the sea between it and the shore to a shallow lagoon. A fringing reef is more like an extension of the land, and in place of a lagoon tends to have flats which may become partly exposed at low tide. An atoll resembles a roughly circular barrier reef, sketchily enclosing a lagoon. Even before the *Beagle* visited the only coral atoll of its voyage Darwin had evolved an elegant theory to explain how a coral atoll might be formed. He thought the process he described would apply equally well to barrier and fringing reefs since although the disposition of nearby land might be different, the principle had to be the same because of the constraints governing the life cycle of coral polyps. He took for

---

[1] Boyle Somerville, *The Chart-Makers* (1928).
[2] Susan Schlee, *A History of Oceanography* (1975).

granted the truth of three assumptions: that corals needed warm shallow water in order to flourish, that parts of the Earth's crust were sinking at a very slow rate, and that corals could grow upwards at least as fast as their substrate sank. The idea must have come to him as a result of finding marine fossils in the high Andes, of realising how portions of the Earth's crust had in the past undergone considerable vertical movement and knowing there was no reason to think this process had stopped. This by itself was a radical idea at the time, as we know from the controversy Charles Lyell's geological theories had provoked.

Darwin proposed a volcanic island poking up out of the sea to whose shallow inshore waters coral larvae drifted, forming colonies. In time these colonies merged and became a fringing reef more or less surrounding the island. The island itself, meanwhile, was sinking and obliged the corals to build upwards so as to remain in shallow water. Since the island was cone shaped it shrunk as it sank, producing an ever-widening lagoon around itself. The innermost corals, now cut off from the constant supply of nutrients in the open sea, grew no more, so that when the peak of the island finally disappeared it left a lagoon roughly surrounded by a ring of coral whose outer edge might extend straight downwards to a great depth as successive generations of polyps had built on their defunct predecessors. In this way, the coral ring of an atoll marked the original outline of a vanished island.

This neat theory of Darwin's was contested, notably by the Scottish oceanographer John Murray. After the sudden death in 1882 of Charles Wyville Thomson, who had led the *Challenger* expedition in the 1870s, Murray took over the job of publishing the expedition's immense report. He had been on the *Challenger* himself and was very struck by the sediment samples which revealed far greater precipitation than he had imagined. He now thought Darwin's theory too complicated and too dependent on crustal movement. Instead, he proposed that the calcareous remains of plankton, falling on undersea mountains in a steady drizzle over the millennia, built up a layer of compacted sediment reaching to within two hundred feet of the surface, at which point the coral larvae started their colonies. In order to explain the atoll's characteristic shape Murray somewhat weakly suggested that this was simply the normal pattern of growth for coral colonies. Trees grew into tree shapes, corals into atolls.

In order to settle the dispute, scientists visited the Great Barrier Reef and the Maldives (where the word 'atoll' came from) at about the turn of the century, but the results of their researches were inconclusive. It was obviously going to be necessary to bore straight down into a reef and take core samples. If these turned out to consist of dead coral, Darwin would be vindicated; if it was sediment, Murray. HMS *Penguin* (without Boyle Somerville on board) sank some holes and the findings seemed to support Darwin, but they were challenged by champions of Murray's theory who claimed the results were false because coral fragments had fallen down the borehole as it was being dug.

The problem was not finally solved until after the defeat of Japan in 1945. Suddenly the United States found itself in possession of hundreds of scattered coral islands and atolls which had formerly belonged to the Japanese Mandate. Needing somewhere isolated to test nuclear weapons, the US searched its new trust territories and hit upon the Marshall Islands. At that point a serious investigation was made of the underlying geology, in the course of which scientists took core samples from deep borings. In Eniwetok and Bikini atolls they went down to over 4,000 feet and found it to be solid coral all the way to the basaltic bedrock, interspersed with strata containing the fossils of land snails and pollen which showed that the island's generally downward progress had been interrupted from time to time by violent upheavals which had brought it back above sea level for a few millennia. Darwin's theory was essentially, if belatedly, proved correct.

Since then it has become clear that many reefs (including the Great Barrier) are formed in ways rather more complicated than Darwin's simple schema. In 1919 the American geologist Reginald Daly proposed that glaciation during ice ages would have had just as much effect on corals in terms of their growth and erosion as geological sinking. Current anxiety over global warming and the projected rise in sea levels has revived Daly's ideas, and in places like the Maldives it should soon become apparent to what extent changes in sea level mediate the formation of coral reefs.

The centuries-long dispute about the nature of coral is rendered neither obsolete nor irrelevant by modern science. True, there is no longer any doubt about the organisms responsible for reef building and—broadly—how they do it. At levels of biochemical detail, however, there is still much to learn. Most reef-building corals, for example, exist in symbiosis with microscopic algae. A single coral polyp looks very like a miniature anemone, its close relative. It has rings of stinging tentacles surrounding a mouth, all of which is able to contract defensively into its stalk. Living within its tissues are the algae which among other things perform photosynthesis and fix nutrients for their host. More than that, the algae enable the polyp to secrete stone. This is a most remarkable attribute and brings a certain accuracy to the old name 'madrepore.' Since algae able to photosynthesise are considered plants there arises the peculiar arrangement whereby a plant and an animal combine to produce a mineral—in this case pure limestone. The chemistry by which this is done is not yet entirely understood. There is great complexity in the way these symbionts interact both with each other and with the nutrients in the seawater, with varying temperature, degrees of salinity (fresh water is fatal to corals, which is why fringing reefs are always broken at river mouths), with currents and with light. Not the least striking part of a coral reef's equivocation, therefore, is that it imprisons at its heart a gigantic plant, while on its surface and slopes there may be few marine plants visible since herbivores such as surgeon fish and sea urchins constantly graze seaweeds back to their roots.

The swimmer among reefs likes to know such things, likes to shine torches into cracks and crevices, takes pleasure in seeing a tube worm sense his presence (smell? sound? shadow?) and retract in a flash, takes pleasure also in knowing the worm ate its tunnel into the coral limestone by secreting acid. It is not only in details that we experience reefs, though, any more than we experience a forest by examining leaves and marvelling at processes of gas exchange. Both reefs and forests may be studied closely but are experienced as environments. So viewed, coral reefs are true borderlands, abounding in all sorts of ambiguity. Many of these ambiguities are set up by the classificatory systems which have been used to make sense of phenomena that refuse their assigned niches. The swimmer who daily goes down among corals to watch and listen soon becomes aware of something in this rich profusion which corresponds to the so-called 'dark matter' postulated by astronomers to account for there not being enough visible matter in the universe to satisfy theory. (In this sense 'dark matter' can be viewed as the astronomical equivalent of the vanishing islands of the North Atlantic.) The 'dark matter' in reefs is the subject of the second part of this chapter; it is enough for now to point out that any form of classification, merely by drawing an imaginary border between two groups of objects, spurts into sudden being a third space as real as the counterbalanced pole which marks the frontier between two countries but is not in either one. Merely to propose that A is not B automatically brings a third coordinate into play, the offspring of wherever it is one stands in order to dispense categories, and which partakes of both A and B. All liminalities belong entirely to the mind, and we are perverse if we expect the objective world to keep to our categories.

At the moment the whole corpus of Linnaean taxonomy is in dispute as radically new ways of grouping creatures are proposed. The way we view the natural world is unconsciously influenced by our received picture of a stylised evolutionary tree which over the past two hundred years has become steadily more ramous and bushy as efforts are made to graft fresh discoveries on to the existing trunk and limbs. Changing this schema would have an interesting effect. There is no serious question that some form of Darwinism is the best model we have of how things evolve, but there is much doubt about traditional notions of the relationships between them. Modern taxonomy has to confront the possibility that the hypothesis of 'Mitochondrial Eve,' the putative mother of the entire human race, may eventually become elevated to a theory by advances in genetics. The category of *species*, the lowest classification of all, is under constant pressure not only from modern fieldwork but from palaeontological evidence like that of the Burgess Shale.[1] Maybe coral polyps will one day find themselves in an entirely new category (or back in an old one) of plant-animals, now based on something like ways of dealing with nitrogen.

Vladimir Nabokov, who was once simultaneously the Curator of Lepidoptera at Harvard and Professor of Literature at Cornell, said of writing that 'the precision of the artist should accompany the passion of the scientist.' This is an admirable recipe for swimming among reefs, where details must be avidly noticed but never seen in isolation. Modern entomology no longer distinguishes between moths and but-

[1] See Stephen Jay Gould, *Wonderful Life* (1990).

terflies (Nabokov's particular interest), the classification having been deemed too arbitrary to be meaningful. All those
Victorian clergymen, earnestly sorting knobbed antennae from
feathered antennae and mildly put out to discover day-flying
moths, were making distinctions which have turned out not to
be taxonomically useful after all; except that no discrimination
is ever wasted in either science or art since all interest derives
ultimately from the ability to spot difference.

A reef project (how full the world is of reef projects!) was
under way a few years ago off the coast of Kuyo Island. At the
time I was living on an islet in the same group. On a visit to
town I met Jim Parkes, who had until recently been a Peace
Corps volunteer, had since gone away and returned with a
technical qualification and was now seconded to the Bureau of
Fisheries. He took me some miles to a dismal scene. Offshore,
was a typical fringing reef which at low tide became a flat,
extending maybe two hundred metres to a white line of surf.
Ordinarily it would have been an expanse of puddles and small
pools at low tide, dotted with boulders and slimy with liver-
coloured blobs of this and that. One would have expected to
see groups of villagers, mainly children, with tins and plastic
bags, going through the tidal pools for edibles: shells, baby
octopus, blennies and gobies. Instead, the reef was a destruction site. A new causeway of blinding white coral chips led out
to the middle where a yellow excavator stood up to its axles in
seawater, fanged bucket plunging and rising, disgorging a rubble of water and corals into the back of a truck.

'We've got a problem here,' said Jim. Up to then I had not
especially liked him but was now won over by this unsuspected
laconic talent, for the scene was one of ruin made ironic by its
backdrop of a tour operator's tropical idyll. A long white beach
stretched in a curve backed by coco palms, all done in the
three primary colours of holiday brochures (blue sea, white

sand, green fronds) whose baldness in print never hints at the possibility of considerable subtlety. 'What we've got is a company mining coral for construction and a fishing community without any fish. And a mayor up for re-election.'

The coral had been mined here for some years before the heavy machinery came. Men with crowbars had levered out chunks which buffalo hauled away on crude sleds. Coral could be crushed and spread as a gravel bed for roads, or roughly dressed could be used for building in place of the ubiquitous hollow blocks. 'Labour's cheap, cement costs,' as Jim remarked. 'Now, the gentleman who owns the rig and the truck is a local who sort of inherited them from a government project to build a road around the island. That failed from lack of funds and equipment, largely because they were all embezzled or stolen in the first month. I've got this congressman friend in Manila who has his own reasons for not wanting the beaches of this island ruined . . . No, you don't want to know about it, believe me . . . So although he doesn't realise it yet, within a week or two this guy with the 'dozer's going to get zapped by Presidential Decree number twenty-eight thousand three hundred and eleven-six which forbids the destruction of the foreshore owing to the fact that it's national property. Problem is, what to do with the remains.'

About half a mile of immediate reef had been destroyed, the topmost four feet having been stripped out. One consequence was that the corals on the surviving outer slope were visibly suffering since the outgoing tide was now bringing with it none of the nutrients and planktonic matter generated by corals inshore. Jim and I swam up the reef and he showed me a further effect of the excavations. There were powerful currents on this coast, in some places running almost parallel to it. Lowering this reef flat by a mere four feet had caused a neighbouring bay to be scoured of its sand which the current was dumping over adjoining reefs, smothering them. Local fisher-

men had been driven on to reefs further along the coast, leading to friction with those already there, to overfishing and the
increasing use of dynamite in an attempt to make up for smaller
catches. This in turn destroyed corals which led . . .

'You get the picture,' said Jim. 'It's bad wherever it happens
but they're doing this sort of thing even in the Maldives, which
has to be cutting their own throats. The highest point in that
country's about two metres above sea level, and since it's all
coral there's nothing else to use for building. Probably the best
thing there would be to earmark an entire atoll for excavation
and ban it everywhere else. As it is, when you remove coral
piecemeal, like here, it's hard to see the end of the knock-on
effect. Last week a fisherman was killed about five miles down
the coast. Dynamite. Got the fuse wrong, took off his arm and
half his head. OK, you can say that was his fault, it was his
choice to go fooling about with gin bottles packed tight with
fertiliser. On the other hand you could say he might never have
done it if it hadn't been for this thieving dickhead here.'

Our swim had revealed something else about the surviving
corals on the seaward slope, which was that they were showing
the first signs of bleaching. This is a condition which happens
when the polyps eject their symbiotic algae. Since it is the algae
which give polyps their colour (true corals produce pure limestone, which is white) the corals blanch, the polyps becoming
no more than a milky slime on their surface. It is not yet known
why bleaching occurs, only that it is probably a sign of stress.
There may be no unitary cause. The polyps can recover spon-
taneously or die, in which case the corals often turn green as
their skeletons are colonised by ordinary algae. It is unsettling
to swim along a reef slope and see among the profusion of
coral species the bone-white beginnings of blight, a marine
leprosy mysterious and as yet patchy. It conveys only that
something is wrong and likely to become worse, that everything is more poised on an edge than one knows.

A few days later I was in the middle of the small island, inside the belt of coconuts and among the paddies and fields of peanuts. Behind the last plantations rose an abrupt low cliff of dark grey rock from which grew creepers and swags of greenery, an escarpment topped with uneven patches of forest. It was evidently from this miniature lost world that each evening the great fruit-eating bats flew out, many of them spanning more than a metre, to descend on the cultivated coastal belt. There they fed on papayas and drank the palm sap being tapped to make toddy. The cliff was no more than one hundred feet high, generally less, and was made of fossil coral. Everywhere were the regular patterns of polyp colonies, the dimplings, pores and stars of tiny stone-secreting animals whose labour had built entire islands as well as the Great Barrier Reef, the biggest single structure on Earth. At some time in the past the flat top of this island where I stood among the tamarinds and cashews had lain beneath a shallow sea rippled with sunlight and thronged with blue devils and fuseliers like chips of sapphire. Having sunk and then risen it was presumably now sinking once more. It was difficult to make sense of such geological eras because the inconceivable slabs of time could briefly be prised apart and wedged open by rudimentary arithmetic. If corals could grow at the rate of an inch a year, which under optimum conditions they can, then this entire escarpment might only have taken 1,200 years to lay down— say from the time of *Beowulf* to the present. A mere nothing, really.

Why, then, could it not be dismantled again? If people needed building materials it made more sense to break up dead coral deposits than to destroy live reefs supporting a complex local ecology which included many hundreds of villagers. I put this point to Jim but he said fossil coral was far harder than recently formed limestone. The outside might appear friable, but that was just weathering; inside it was like

granite. I did not know if this was true. At any rate the idea
came to nothing and one was left with the sense of a gross
imbalance. It seemed abysmally crude to smash up a construction of microscopic architecture in order to build other structures of a coarseness grotesque in comparison. What, then, was
Jim's 'project' if not merely to stop the mining of coral?

'Re-establishing fish, naturally,' he said. 'Or unnaturally.'

The best thing to do was simply to leave the hacked substrate alone for a century or two, by which time and with any
luck new corals would have established themselves. Sure, said
Jim, the best thing so long as you had an alternative source of
food. Hadn't I got the idea *yet*? This business wasn't particularly to do with corals, it was to do with fish and, eventually,
with votes. If I tried thinking of the fishermen as farmers, then
these reefs were their fields. Plough up a field without re-sow-
ing it and you'd get nothing but a crop of weeds.

I supposed so. The truth is, remedial work bores me. If one
is a fatalist one believes that once something needs to be
restored, to be caught and fought for on the edge of extinction,
then in a sense it is already gone, has already been lost in the
form that had meaning. It is a pleasing irony that any revival-
ism, like life support, always suggests demise and not survival.
Why should this be? And does it mean that someone who
recovers after terrible injuries and weeks of intensive care is no
longer authentic? At any rate, Jim's solution was to create an
artificial reef out of old truck tyres. He was going to take deliv-
ery any day now of several hundred tons and dump them all
over the ruined reef. It was, he said, an experiment which
would be watched closely in many parts of the world. If it
worked it would solve two major problems at once, since up to
now nobody had found any use for old tyres except maybe as
fenders on wharves. Tyres would make terrific habitats for reef
fish because if dumped higgledy-piggledy they formed an elaborate honeycomb of holes and crannies for shelter. After a time

they would become encrusted with marine growths and maybe eventually even corals would take root and start building. In the meantime they would be colonised by fish.

By now I had grown quite fond of him despite his stainless-steel faith in remedies and projects and even—quite possibly—in progress, too. But away from Jim I found myself caring only slightly that well-meaning, busy people like him should succeed in their tasks. I knew all about the fishermen but saw no reason for not holding contradictory views at one and the same time. It is not possible to balance an equation between aesthetics and somebody's livelihood. They have nothing to say to each other. As far as I was concerned the dumping of a shipload of old lorry tyres into a man-made gap in a fringing reef was adding insult to injury. If the equation were to be solved equitably it would entail the agents of the damage paying its victims compensation for as long as it took the corals and fish to re-establish themselves. Jim said I was a drivelling idealist, flitting about the world like a disdainful butterfly irresistibly attracted to decay, content merely to feed off it rather than do something about it. I warmed to him still further, saying he was an excellent judge of character. Butterflies do not believe in the efficacy of *doing*, but they definitely are fond of flitting and feeding and drivelling. We bought each other a series of drinks until the Mayor who would be mayor again came in and began a lengthy discussion of *arrastre* or lighterage fees for unloading hundreds of tons of perished rubber.

II

Sensing the Oblique

J ust as law courts in certain hot countries provide a habitat for diverse species of people, so does a reef supply the architecture and ambience for all manner of marine predators and prey, hangers-on, refugees and indigent waiters on scraps. It has its diurnal and nocturnal rhythms but the activity never entirely stops. After a strange twenty-minute hiatus, night creatures follow day creatures through the self-same corridors and chambers and are themselves distinct; but unseen and in no particular place long, long processes continue imperceptibly. Even in the depths of night scattered lights wink in the windows of these unsleeping cities.

As a ramshackle social nexus a tropical law court is awash with voices. The sounds seem almost generative, as if every voice secreted a molecule of stone and buildings grew out of pure speech. Instead of some general administrative order, it is the uproar itself which supports and enlarges the monumental fabric. Around its walls and beneath the shade of banyan trees sit men at tables with ancient typewriters, patiently transcribing voices from the throng: aggrieved, supplicatory, self-righteous, wheedling. For every sentence typed a thousand ascend like smoke past the birds in the banyans. Hot knots of waiting peti-tioners discover each other and put their cases over and over again, ever more articulate, ever more impassioned, scattering in the sunlight gems of rhetoric whose brilliance goes unrecord-ed. Squeezing through the press of people the tea boys come and go with battered tin trays and dirty glasses with half an inch of undissolved sugar in them. Wholly unconcerned with legal wranglings they form a noisy sub-community of their own, their cries and whistles joining the birds. Everywhere is a prodigality of speech and gesture and smell, and from it all a distillation leaks out in a steady trickle of files containing depositions typed shakily on yellowed paper.

An old man wearing spectacles mended with fuse wire gathers them up every so often and takes them inside the great building, through a back entrance and gloomy passageways, climbing flights of worn steps until, having knocked on a door, he enters a remarkable room. It is not particularly big but the open windows are spacious and it is the light coming through them which gives the room its quality. This must be the back of the building, facing away from the hubbub downstairs which had seemed to besiege it on all sides. The air is quiet. The windows open into a cage of leaves, as if the room were built in the heart of a tree. Through them filters a serene undersea glow pricked with spicules of dazzling sunlight, one stray beam laying the spectrum in a bar of colours across the window sill. At a bare table before the window sits a scholarly man—a judge, possibly—staring out into the greenery and watching the hop and flitter of small birds among the branches. The old messenger lays the files in silence on the judge's desk and in silence withdraws, closing the door gently behind him. Maybe something of the din below is audible after all for there is a faint but constant background noise, a soft roaring like a distant sea. The judge sighs, opens the top file and begins to read. It is hard to connect the orderly, formulaic sentences with the whizzing and tumultuous lives which, in some garbled form, they partly express.

One needs to drift in the green undersea light day after day, month after month, maybe for years, until almost bored. Or maybe not bored but blocked. Looking and listening, the conviction grows that something is missing, some dark matter hidden but deduced. It is an absence which in an animal privileging the visual leads to the idea that it must be *located*, if only it could be found. Somewhere behind the next outcrop (vaguely right-angled, like the wing of a sombre old palace) must be a forgotten courtyard containing a great assemblage, a tumult, the core and centre of this submarine complex. Until it is found one will go on missing the point, ears stuffed with water and eyes straining behind glass, too immured in a scholarly attention to detail and too intent on deciphering easily apprehended messages which purport to tell the whole story. What is needed is a sideways shift, a skidding off into a different position entirely.

Some black art may be needed for dealing with reefs, one I have never discovered. All my experiments with them, while having suggested themselves as serious, ended up as parodic or whimsical. When diving off coral reefs it would be useful to learn how to judge the direction and speed of a nearby boat rather than run the risk of surfacing for air directly in its path. Even after so many years I still cannot do this reliably. Sound under water becomes omnidirectional, reflected off rocks, off the shelving sea floor, even back down from the surface. Depth, too, makes a difference. The more distant a boat is the further down one has to go to hear it. Sometimes a fishing boat will pass directly above. The silver-edged lozenge scoots overhead—black, very sharp and swift—and the whirring disc of its propeller mutes itself for an instant, like an electric fan which is quieter heard edgeways on than from in front or behind. What, then, would be the effect of hearing a sound from a source vertically below instead of overhead?

Having borrowed a cassette player with a loudspeaker I swathed it in many plastic bags and walked into the sea, its owner watching and torn between consternation and ridicule. I swam down about twenty feet and set the machine on a ledge among corals. The pressure flattened the layers of polythene in

a secondary diaphragm across the speaker grille and it was possible to believe equally that this might conduct the sound better or else muffle it. I wanted to see what effect music had on
fish and other reef creatures, but also how well it would sound
under water. It turned out to be disappointing, ethereal only in
so far as it was inherently weird to hang in water twenty feet
down and hear Mozart's G minor Quintet coming from behind
an outcrop of *Acropora*. It was not so much attenuated as
muted, the higher frequencies suffering most. With the cassette player on its back the sound, as heard from the surface
directly above, was definitely feebler than at the same distance
away horizontally. This was probably due to the focusing
effects of rocks and the sea floor. As for the creatures, they
paid the music no obvious heed, with the exception of a
damsel fish which braved the ecstatically depressed sounds to
dart at the polythene. They are highly territorial, and the player was probably in its back yard. Since fish have excellent hearing it is likely that this dim source of noise was at all the wrong
frequencies to be interesting. It certainly was muffled. This
maybe had something to do with water pressure deadening the
air column inside the speaker and even slightly inhibiting the
vibrations of the paper cone itself. Altogether, it was a pathetic substitute for a GLORIA transponder.

On other days I experimented with making high-frequency
noises by partially inflating condoms and trying to make them
squeak under water. It was necessary to wash off the silicone
lubricant otherwise my fingers would skid greasily without
making a sound. I finally coated my hands with a resin used
locally as the basis of incense and tried inflating the bladders
to different dimensions to obtain different frequencies. This,
too, was a failure. At best I achieved a brief grunting sound,
though it was uncertain whether this was generated manually.
Anyone who has ever tried swimming underwater holding an
inflated condom will appreciate the difficulties involved.

There are fish which are blind and others which have no olfactory organs, but they all have some variety of acoustico-lateralis system. That is, they have 'ears' or sound receptors as well as lateral line organs. A fish's lateral line runs from its head along both sides, its course often marked with a pigmented stripe. It is made up of tiny hairlike sensors which respond to changes in the water caused by local movement and currents. (It was probably a similar mechanism which triggered the tube worm.) In addition to having 'ears' some species of fish have the capacity to make sounds, generally in one of two ways. They either drum on the wall of their swim bladder or, like a cricket, stridulate two bony surfaces together. It is presumed that such abilities are used for courting and mating rituals. Species which have some sort of connection between their swim bladder and their inner ear must have exceptionally acute hearing since the air sac would act like a diaphragm and efficiently collect sound. As for the noises made by marine creatures, some are very loud indeed. Apart from the celebrated carrying qualities of cetacean sound, the sharp snapping of a pistol shrimp is enough to make a swimmer jump, while the grunts of toadfish have been known to set off acoustic mines.

Such things are the background to another experiment, one performed so often as to become a habit. This is to swim out to the edge of the reef on a moonless night, head down into the depths and, holding on to a rock near the bottom, simply concentrate on all that can be heard and seen until the air in the lungs runs out. This is the best way I have yet discovered of apprehending a reef and it has become the central ritual of my explorations. To save blundering painfully into corals, stinging hydroids and sea urchins one needs to take a sealed torch, but it must be used sparingly otherwise the light destroys the eyes' rhodopsin and leaves one blinder and—strangely—deafer as well. Obviously it is better to choose a reef whose layout and fauna are familiar. Knowing what to expect it is less of a shock when a stingray explodes off the bottom in a cloud of silt, pelting wings flashing quick beats of white underside as it vanishes into the blackness. It is necessary to be alone and it is essential to be apprehensive. When the night is overcast and the wide drench of tropical starlight falls uselessly into an upper bed of soft cloud and nothing below it can be seen, not even the shore then is the time to take a deep breath and swim down, grains of luminescence streaming back from fingertips, down and down.

It was whale song which mariners heard filtering through their vessels' resonant wooden hulls and which they took for the Sirens' voices, beckoning them to disaster. To have lain in one's bunk at night and heard on the other side of a few inches of oak and copper sheathing those directionless, distanceless cries must have been to feel the chill of utter melancholy and dissolution—also, to have felt one's nakedness. This is the effect of listening to reef sounds at night, too. It is more than just the nakedness of wearing next to nothing, and it is more than vulnerability. It is the sensation of animal messages passing *through* one as if, being seven-tenths water, one's body were transparent.

At first it is too unnerving to permit concentration. After a time when nothing life-threatening has happened the rhythm of swimming down, waiting, and coming back up for air becomes soothing. The sea is warm. With the clamp of water over one's ears and the blackness pressing up against one's mask conditions approximate very slightly to those of controlled sensory deprivation, a disorienting and eventually unhingeing technique fashionable in torture and interrogation circles some years ago. But there is no real comparison. There is too much sensation, too much physical effort in holding the breath, in staying down rather than floating up, in seeing and hearing. It is never more than mildly hypnotic for a few minutes (but with a vanishing of time which makes those moments impossible to calculate). Steady in the background is the loud white noise of uncounted crustaceans stridulating with pincers, horny plates, mandibles, who knows what. Very occasionally it stops dead, and in the ensuing silence a chill passes over the body because a million crabs and prawns have all heard something attention-grabbing which one has missed completely. What is it out there? Out here? The frogs in the paddies do it, too. Nightly they crank up their ranarian machine until it is turning over at a constant speed. It goes for an hour, then abruptly stops. A beat or two, then a few brave ones try to turn it over, are joined by others until it catches and settles back into its rackety tick-over. Crickets will do the same.

Out here we are on the edge of something: of drowning, fear, understanding. The huge unseen city itself seems always on the cusp of vanishing, it is so delicate and its true nature so elusive. It is a place whose strangeness is far greater than we can know even as we painstakingly try to identify each snapping shrimp, each grunting fish, the soft concussion—like a cloth being flapped—of a sizeable fish taking evasive action somewhere nearby. *Whup*. But then we, too, are stranger than we imagine. We hang here in the depths with granules of cold fire prickling around us as creatures and currents stir dinoflagellates into luminescence. We hang here, inquisitive carbon-based life forms, knowing that every atom of carbon now in our bodies was once in the interior of a star. For an instant, we dissolve, are without form, become nothing but the point at which the three axes plotting this three-dimensional borderland intersect. The three dimensions of a fringing coral reef are as follows. *Horizontally*, it marks a border between sea and land. The flats at low tide baking for hours in sunlight support a variety of marine animals which can survive out of water and resist a wide temperature range. *Vertically*, the seaward cliff of a reef face reaches to within inches of the surface and may plunge 1,000 feet in a precipitous slope, its initial descent marked by a steady change in colours and life forms. *Obliquely*, a reef exists at other wavelengths than those we can perceive. The hidden courtyard with its tumult, the babble and rhetoric, the colours and unimagined smells, all are tucked away in great bright pockets of the electromagnetic spectrum which are closed off to us, in sounds we cannot hear, in pheromones our nostrils cannot detect. This knowledge makes us ache, sea creatures that we once were, as for a country we have lost on the far side of a frontier we can barely even discern. We are left with our narrow, thickened senses. We are also left resorting to fossil gestures. I find myself opening my mouth under water, the better to hear. It works on land, but not noticeably beneath the sea, where a proportion of sound reaches one via direct conduction through the bones of the skull. It is the gesture of a creature with a phylogenetic memory, as if some ancestor with a different otopharyngeal arrangement had opened its mouth to let messages in and out, grinding chattily away with a set of bones at the back of its throat. I also track down a dimly surviving fragment of my lost lateral line, maybe. In daylight one day I notice tiny plumes of silt pulsing from a hole at the foot of a coral outcrop about thirty feet down. I do not know what particular worm or shrimp is busily excavating inside but have the sudden urge to feel this minute puffing as well as to watch it. I put my fingertips half an inch away but can detect nothing. Without thinking I manoeuvre awkwardly so that my head is close to the hole. The puffing stops for a few seconds but then starts again. Gently I move so that I am almost kissing the hole and can just feel the tiny waves of energy from the unseen creature's fins or pedipalps break against my upper lip. It is the underwater version of that unconscious gesture which makes people press their washing beneath their noses to feel how dry it is: ex-weanlings whose haptic sense for moisture and movement was designed for the breast.

The reef's vertical axis is most vividly revealed in terms of light, ranging from the brilliance of sunlight to the inkiest depths. The sea is both lit and heated by the sun's energy which is absorbed and scattered from the moment it penetrates the surface. The uppermost metre of the sea effectively absorbs all ultraviolet and infrared, respectively those wavelengths shorter and longer than the visible. Thereafter, seawater absorbs the longer wavelengths first and at about thirty feet down most of the surviving energy is in the blue-green part of the spectrum. From here downwards, increasing numbers of reef creatures are coloured in various shades of red. With almost no visible red light remaining they look dark or black and in still dimmer waters further down become almost invisible. In these top ten metres the simplest experiments show how much light the water absorbs. Slightly dull objects take on fabulous colours as one swims towards the surface with them. Likewise, a blood-coloured anemone becomes pale and anaemic as one swims away from it.

At a depth of fifty metres only five per cent of the sun's energy still penetrates. If the water is exceedingly clear there is enough light for photosynthesis in plants and algae down to 150 metres. Below that, ordinary plant life cannot exist, which is why there are no great prairies of seaweed covering the deep ocean beds.[1] Generally speaking, the 100-metre mark defines

[1] This statement wants qualifying. Plant-like structures may occur in the deep and highly specialised ecosystems of vent communities. These grow up around 'black smokers' at volcanic sites and are not based on photosynthesis. The bacteria of vent communities, on which populations of tube worms, huge crabs and oysters and other creatures depend, have metabolisms which use sulphur in place of oxygen. Hence these strange pockets of life have nothing to do with the upper world and its biochemistry. Meanwhile, the greatest depth at which conventional plant life has been found is 269 metres, a clump of maroon algae in exceptionally transparent water off the Bahamas. There is a group of cold water corals that form limited clumps and reefs at depths between 100 and 3,000 metres and at temperatures from 4° to 8°C. These are the azooxanthellate corals such as *Lophelia pertusa* whose outcrops off Norway and the Shetland Islands are often ravaged by deep trawling and oil and gas prospecting. These corals lack symbiotic algae, so have no need of light for photosynthesis. Their polyps depend entirely on their tentacles for food and consequently the corals are slower growing. Although individual outcrops are more limited in size than the zooxanthellate corals, they form habitats for a rich biodiversity, especially of invertebrate species.

the bottom of the euphotic ('well lit') zone, the most productive part of the sea. It was also long believed to mark a theoretical limit for unaided human divers. The 'constant weight' diving record (i.e. with held breath but without fins) is currently 80 metres (105 metres with fins), although the 'no-limits' apnea record was broken in 2005 at 209 metres. Such extreme feats are made possible by the mammalian diving reflex, itself highly suggestive of an aquatic past for *Homo*.[1] In splendid violation of a supposed boundary auks (razorbills and guillemots) have sometimes been seen from submersibles at one hundred metres while the bird diving record is easily held by the emperor penguin with a recorded 18-minute dive to 265 metres. In any case this 100-metre zone effectively denotes the end of the reef as an ecosystem. Below this its dead corals become a habitat for scattered sponges and, of course, the sundry twilight animal species.

Underwater photography naturally has to take into account this absorption of light energy by water. Even if there is enough ambient light for filming, colour values will be increasingly affected by the progressive filtering out of the warm and comforting wavelengths: first red, then orange, then yellow. 'Correct' colour values can be restored by carefully calculated artificial lighting. Any approach to the question of human vision under water leads back to the fact that the narrow waveband in which our eyes operate corresponds very markedly with how sunlight is transmitted in seawater. Quentin Huggett at IOS has wondered whether this setting of our visual 'window' reflects our aquatic origin. In air the bandspread is much wider. Several species of insect see at ultraviolet wavelengths,

---

[1] The mammalian diving reflex is triggered in humans when the face is immersed in cold water. This immediately reduces the heart rate by up to 50 per cent, restricts the blood flow to the extremities in favour of supplying the brain and heart, and shifts blood to the thoracic cavity and lungs. This prevents the lungs collapsing under pressure, which otherwise would happen below about thirty metres.

while the pit viper uses infrared sensing to detect warm prey in the dark. Had *Homo* developed differently he might have 'seen' a different world.

It is sad that we cannot smell things underwater, though now and then something lodges in a taste bud or receptor to produce the simulacrum of a smell, a pungent impression located somewhere in the muzzle part of the face, neither precisely smell nor taste. Sad, too, that our hearing is not very acute, and with a limited range from about sixteen cycles per second up to 20,000 cycles per second. A mere cat has three times this range, a bat six. The pitch at which human ears are at their most sensitive is that of 'a child's or woman's cry,' according to Yi-fu Tuan, who thinks this suggests that our ear is adapted to favour our species' survival rather than hunting.[1] This is most aggravating to those reef haunters who have no wish to listen to children and women crying, either above or below water, nor any interest in the survival of their own species, but who would dearly like to eavesdrop among the courthouses, malls, temples, palaces and suburbs of their passion.

My lame experiments sound foolish—*are* foolish to scientists like those aboard *Farnella* who could suggest any amount of gear for increasing my sensing ability. Yet in the early days of submarine warfare the help of people with musical knowledge and perfect pitch was sought in order to classify the sounds made by submerged craft. In World War I the composer and conductor Sir Hamilton Harty was called in by the British Admiralty's Board for Invention and Research to identify the most likely frequency bands of hull and propeller noises, 'anticipating by a whole war a similar attempt in America, where the conductor André Kostelanetz was approached for much the same purpose . . .'.[2] Ernest Rutherford also took a colleague with perfect pitch out in a small boat as part of the

---

[1] Yi-fu Tuan, *Topophilia* (1974).
[2] R. V. Jones, *Reflections on Intelligence* (1989).

war effort. At a prearranged spot, one of the great names in atomic physics took a firm grip of his companion's ankles while this man stuck his head into the Firth of Forth and listened to the engine note of a British submarine. Hauled back into the dinghy and towelling his head he announced it was a submersible in A flat and he would recognise it anywhere.[1]

Today, with an ultrasound detector I could wiretap the citizens of this submarine city, but it would not be the same. As with medical diagnostic equipment, such apparatus can also blunt other responses. Ironically, part of the problem of the tropical reef is its very *visibility*. It famously stands as the icon of marine exoticism: brilliant colours, profusion of species, intricacy of shape and design. To make all this more easily seen by means of scuba gear, or more easily heard with ingenious electronics, is somehow too facile to be serious, except for detailed scientific work. Scuba equipment is, of course, indispensible for any work below forty or fifty feet and down to two hundred feet. It is a marvellous invention.[2] However, it has certain disadvantages which are not compatible with travelling alone in remote parts of the world since it is heavy to lug around, expensive to replace when stolen and limited by the local availability of a reliable compressor, to say nothing of the obligatory diving companion. Apart from wishing to be quite alone in the sea, I dislike scuba gear for two additional reasons.

---

[1] See *New Scientist*, 1768, p. 80.

[2] Scuba gear is a vast improvement on all its predecessors, although the tonic effects of oxygen seem not to be as pronounced nowadays as they used to be. In a report in the New York *Daily Times* (24 August 1854) the writer considers the newly developed diving suits of the day which were made of leather and rubber and entailed the divers carrying 'a box of condensed air.' 'The condensed air they are forced to breathe, furnishes them a greater quantity of oxygen in a given time, and increases their strength very much for the time being. A diver, at a depth of ninety feet under water, at Portsmouth, England, was known to bend nearly double an iron crowbar in his work, which resisted the strength of four men at the surface.' Presumably the proposition here is that if oxygen gives life, a lot of oxygen will give a lot of life.

It is awkward to wear, all the time making me conscious of itself, of bits of metal and tube, of tanks as bulky as a growth, of belts and dials and rubber straps and harness. But even more importantly, I cannot hear with it. Breaths and bubbles rattle and roar in the ears; the very heart drums in the air hose. It all gets in the way as much as it facilitates, and on most occasions it is preferable to sacrifice depth for the immediacy and greater effort of free diving. The immediacy is whatever transcends discomfort and inconvenience, leaving one uncluttered on borrowed time. By making certain things too easy, scuba equipment gives rise to that curious paradox: the more accessible a thing becomes, the harder it can be to see.[1]

Besides, we are not trying to push outwards the frontiers of science, but our own. We are content to have identified the calm room with the big windows letting in the blue-green undersea light and to know that whatever we learn is only a part of what goes on in its corridors and undercrofts and courtyards. It is a consoling pleasure to have hung about this cityscape at night and watched the lights winking to codes unguessed at, to the roar of conversation in a universe off at an angle. It is enough to have clung to the roots of a reef at sunrise and watched the dark gradually bleach from the water. The night shift changes, silence falls. Then the creatures of day emerge and all around the dawn chorus starts, as if invisible behind still-misty water, a great banyan were spreading its branches.

[1] c.f. Wallace Steven's poem 'The Creations of Sound,' in which he chides a writer (probably T. S. Eliot) whose poems 'do not make the visible a little hard / To see . . .'.

III

MARGINALIA:
Souvenirs

Tourist markets laden with trinkets carry their own charge of melancholy, while seeing the *objets'* provenance can be brutally sad. People in remote places labour long and hard to live by this trade. The sacks and bundles pile up in bamboo sheds or beneath a covering of dead palm fronds: the shells and corals dragged from the seabed and rotted out in heaps; the hardwood cigarette boxes stacked in piles; the sharks harvested for flesh and teeth, jaws and fins; the alligators stuffed; the baby porcupine fish inflated and lacquered; the aquarium fish poisoned.

Often it seems the more that people become urbanised, the more they want about them talismans of nature on their walls, their shelves, their keyrings. Many souvenirs are marks of pilgrimage, like religious relics, and denote travel. Many of these talismans come from the sea. They are tokens of lineage and are to *Homo* what a family crest is to an aristocrat. The blood line lives on. Yet perversely, this importing from one universe to another, from water to air, is invariably fatal. Nothing looks as dead as a seashell in suburbia, a piece of coral skeleton or a stuffed fish. The otherness of a shark's tooth, like that of a fossil, begins to ebb the moment it is held in the hand. *Objets trouvés* should be marvelled at, then allowed to become *perdus* at once. Only thus can the transient pleasure of crossed trajectories be sustained in the memory.

The tourist's trinkets and the traveller's memorabilia so swiftly decay from prized objects into junk because they are never what they were believed to be. As pieces chipped from nature they have a status oddly close to that of a monument. Just as a monument purports to refer to the past but is always contemporary,[1] so the tourist's relic constantly rewrites his version of its origin or the moment he acquired it. The more it tells him of his former self, the more silent it remains about its own past. A shark's tooth is to the living actuality of the fish which once ate with it as a holiday photograph is to the scene it depicts. A wave washes over us as we hold them but it is not—as we imagine—a wave of cheerful or tearful recall. Such trinkets commemorate a moment not of acquisition but of loss. The tooth, the coral, the hardwood, the fish, were all wrenched from life so we can later discard them as the impedimenta of a previous self. From long experience we know this in advance, remembering the fate of souvenirs from former holidays even as we tell ourselves that this time they are authentic. But the things are already dead when we buy them; and their rattling into the dustbin later is the sound of our own hollow attempt to seize the ever-vanishing present.

In this way, our own deaths prey on the deaths of others. All mementoes immediately turn into species of fossils. The intricate tracery of a coral fan, torn last year from a seabed off Barbados and bleached for export, stands as mutely on a mantelpiece as the ribs of any Carboniferous leaf embossed on a piece of coal. Our psychic recognition that last summer might as well be ninety million years ago robes our trinkets in their characteristic sadness. Acts of memory are incapacitating, trapped as we are between *souvenir* and *memento mori*. When the objects have collected enough admiration and dust to force this recognition upon us, we throw them away or donate them convulsively to a museum.

---

[1] This idea was developed by Mark Cousins in a lecture at the Architectural Association in London in March 1991.

But there is no reef after all. The swimmer pursued the dark shadow in the water, counting his strokes, until he realised it was receding at every stroke. He gave up and swam in what he hoped was the reverse direction until now he thinks he has returned to his original position.

Still no boat. It has made no effort to materialise during his brief absence. Maybe there never was one. Maybe he has been out here in this radiant deep for days, even weeks? Only the rope's empty tugging at his ankle as he swam reminded him of how recently he had been anchored to his precious craft.

Surely the sun is lower than it was? For the first time he considers what it will mean to spend the night out here. He is not afraid of the dark, nor is he afraid of the sea at night. Yet out here the sharks are maybe not so timorous as they are in the shallower waters near a fringing reef. They like the freedom of a good depth beneath their bellies. Since he will be neither kicking nor struggling he has hopes of not attracting their attention. Like certain predatory humans, they are beautifully attuned to the sounds of distress. No; he will hang here quietly, counting the stars or trying to see the lights of land. Perhaps it will be impossible to resist peering into the Stygian, sparkling gulf he treads, even though the sight of a great swirl of luminescence turning suddenly in his direction might well herald his end. No doubt the strike will be shockingly abrupt, but for some reason he is more frightened by the thought of a first inquisitive tugging at the painter hanging beneath him like a bell-pull. He does not want to hear his own terror pealing out across an empty ocean. At all costs he will be stoical. Fishermen say shark attacks are painless. One feels great bangings and slammings but no actual pain.

This is stupid. It is still broad daylight and the boat has only just gone. There is no possible way in which it could have drifted more than fifty or one hundred yards. A hundred and fifty at the outside. Well, he will not wait passively in the water until death in one form or another heaves up alongside to gulp him down. He will now try to swim a grid pattern. It will not be easy without any fixable starting point, but nevertheless he will try.

He heads for the sun, fifty strokes, then turns at what he calculates to be 90° and swims ten more; then another 90° and back parallel to his original course. He is pleased to find the surface of the water undulates more than he had thought. This convinces him that the chances of happening on the boat are excellent. No sooner has he thought this than, halfway along the new leg of his search pattern, he spies its prow. There is no question. No piece of water was ever that shape, black and slender and curved like a beak. With a cry he abandons the stupid grid and heads for it. Thank Christ, and about time too. That was close. God, that was close. Never, ever again will he do anything so damned stupidly careless . . .

There is nothing. No prow, no boat, not even a length of floating timber. Nothing but empty water taking its shape from him. His grid is broken and lost. He will never find his original position now. He turns and turns in despair and dejection, incredulous to think he will not be seen ever again while the bloody boat will probably be wrecked on some inshore reef to gladden a poor beachcomber looking for useful spars and panels of marine plywood.

4

Wrecks and Death

A death at sea.

I

M any must have had the fantasy of the sea vanishing and leaving its bed open to inspection, though the imaginative leap may be accompanied by uneasiness about the sea's sudden return, catching the fantasist out in the middle, bent over something of interest. (The ominously rising breeze, the horrid realisation that the low strip of dark cloud on the horizon is actually an onrushing wall of water hundreds of feet high.) Treasure hunters, especially, must all have entertained the fancy of strolling offshore among the wrecks and hulks drying in the sunshine, scattered as far as the eye can reach. The image is likely to be sanitised, based on pictures of stranded fishing boats canted over in the deserts of Kazakhstan where the retreating waters of the ruined Aral Sea have left them. The reality would be one of draped and stenching putrefaction.

Wrecks have a particular fascination because they act as foci for so many preoccupations: death, loss, things being hidden and disappearing, things being discovered and reappearing, hoards of wealth. Irony is added in that wrecks and their contents are frequently quite close to their searchers. They may be no more than a few hundred feet away but the marine universe into which they have passed makes them as inaccessible as if they were miles distant or on the Moon. If one were to classify them, wrecks would conveniently fit into four categories: tombs, time capsules, gold mines and time bombs.

On Pier 40 of Honolulu Port, from which the *Farnella* sailed, stood a huge warehouse, practically empty. I learned from the security guard that until recently it had been used as a dump for impounded contraband and, indeed, below the wharf on the side opposite the ship were moored various oceangoing yachts and catamarans, all of which had belonged to drug runners now behind bars. Their craft were waiting to be auctioned off by US Customs. Before that, the warehouse had contained one of the largest collections of aluminium coffins in the world.

Honolulu, and especially Waikiki, strikes many visitors as moronic in its empty-headedness. It is exactly the sort of place disgraced dictators would choose to go into exile, live in a tropical wedding cake of a villa and die protesting their innocence, all of which ex-President Marcos did. But the island of Oahu has long had a grimmer aspect, one which runs reassuringly beneath the sun-and-surf package tourism. Only a few miles along the coast from Waikiki is Pearl Harbor, while the air base of Wheeler Field at the island's centre has been associated with death ever since World War II. For nearly forty years it was the stopover for American KIAs from the Philippines, the Marshalls and Solomons, Japan, Korea, Vietnam and the rest of Indochina. Those killed in action in every Pacific and South-East Asian theatre of war passed through Oahu, and it was here that bodies returned by the Vietnamese government came for identification. Hence the coffins. The bulk of them are now stored elsewhere; but one hundred yards away from the warehouse, on the far side of rows of dusty new Japanese cars awaiting Customs clearance, was a barbed-wire compound and in this were still a few stacks of the silver boxes, slightly corroded from the salt air and with tufts of weeds growing round the bottom.

It is a positive relief for the visitor to Honolulu to be reminded of war, of anything serious and historic to set against a local culture which is so aggressively frivolous. The dignity of the USS *Arizona* Memorial at Pearl Harbor is a profound contrast. The *Arizona*, which had served with the Atlantic Fleet towards the end of World War I, was one of eight battleships sunk or damaged in the surprise Japanese air attack on 7 December, 1941. When the attack began a few minutes before eight on a Sunday morning the *Arizona* had aboard 1,447 of her full complement of officers and men. When she sank, 1,177 had died on the ship, including the division commander and the ship's captain. Some eight hundred are still there. It was decided to turn her into a National War Grave together with a second battleship, the USS *Utah*, which lies half submerged on the other side of Ford Island from the *Arizona* and still has fifty-eight bodies on board.

The memorial attracts something like 1.5 million visitors a year. What might have degenerated into just another attraction is remarkably sombre. The effect is achieved by an organisation which suggests that visitors have temporarily passed out of the indulgent cajolery of civilian guided tours and into the gaze of military discipline. One is ushered into a cinema by a trooper who introduces a short film with some background information about what was happening in the world in 1941. The film then shows the attack itself, grainy, jumpy, tilting pictures of hectic black-and-white action intercut with modern underwater footage of the ship as she is now: rusty chains and barnacled capstans. At the end, suitably subdued, the group is led out through another door straight on to a wharf and aboard a launch which heads out towards the *Arizona*'s resting place. In this way no ordinary visitor can see the ship without first entering its solemn context.

From a distance one sees the white, perforated memorial building, half bridge and half observation platform, which spans the sunken ship. The guide emphasises that at no point does this building touch the *Arizona*. (It was clear from the underwater footage that the ship may only be looked at, never touched. No gloved diver's hand had reached out to wipe away rust from a hatchway or algae from a porthole.) If anything, this launch trip intensifies the solemnity, not least because the tourist's prerogative of continuous smoking and eating and drinking is forbidden. Once at the memorial the group wanders reflectively around, not talking very much. The older they are, the quieter. There are the ship's bell and a marble wall inscribed with all the names of the dead (as always, men may be sent carelessly to their deaths in wartime but their names are meticulously recorded); yet it is the sunken ship herself which commands attention. The memorial straddles the hulk and on both sides the ship tapers away, its ends marked by distant orange buoys. A few chunks of corroded steel poke up above the surface, notably a great circular drum, the barbette of a gun turret. Otherwise, the *Arizona* remains shadowy, bluish, submerged. Schools of damsel fish nose around the coral growths which have taken hold on her decks, looking for plankton. They are familiar black-and-yellow-striped sergeant majors, though I thought I saw another variety as well, *Abudefduf.* The flitting of these creatures between the observers' eyes and the object of their reflection did not have the same effect as of pigeons circling a cenotaph. They were not tokens of a natural world blithely indifferent to human pieties, but drew attention to the medium into which the victims had passed. The skeletons, the events of 1941 and the fish now inhabited the same world, no part of which had been retrieved for the redemption of daylight and the upper air. We who stood looking down through the fish arranged ourselves in the relaxed, slightly unfocused attitudes of those who musingly watch golden carp in a pond—a quite different posture from the stiff upward gaze of someone confronting a monument.

And herein lies the USS *Arizona*'s unique effect. We are accustomed to look downwards at gravestones but never downwards at public monuments. This sunken battleship is probably the only example of a monument which is viewed from above. The bowed head is at once a gesture of private grief, public respect and national mourning.

On the journey back in the launch our escort told us that the slight oil slick we might notice came from the two or so gallons of fuel oil which still leak daily from the *Arizona*. Legend has it that it will stop seeping on the day the last survivor from the ship is buried. 'I guess the ship is weeping,' said a fellow traveller on the bench next to me. He was a man in his fifties with a moustache, who told me he had come because he had been a boy at the time of Pearl Harbor and remembered it partly for the emotion and partly because no one at home in Connecticut had known where it was. He had not expected to be so moved, he said, but he retained a sound middle-aged asperity, remarking on the irony of our guide's homeward commentary being full of platitudes about peace, friendship and the lessons of war when only the previous day the UN Security Council had passed a resolution which President Bush was interpreting as giving him leave to go to war in the Persian Gulf.[1]

Back on land I visited the nearby submarine museum. Dotted around outside on concrete plinths and pedestals was a display of missiles, all of which looked oddly small and rudimentary. There were very few visitors and the place had a pleasantly mournful air. There was also a large black 'Kaiten'-class Japanese one-man suicide torpedo. Apparently this was never a successful weapon, proving temperamental and difficult to control. Essentially a huge bomb with a little seat in it, it sat on its cement bed and bled rust from rivet heads. A notice said that one successful 'Kaiten' pilot had gone to his death wearing a white bandanna and with the urn containing the ashes of his friend killed in training jammed into the cockpit with him.

[1] I was referring here to President George Bush and the Gulf War of 1991.

The centrepiece of this display was still afloat: a submarine moored to the quay, USS *Bowfin*. She last sailed in training in the early 1970s and the interior looked as though it had been kept polished by use as much as for exhibition. She had survived the last war with a distinguished record of 'kills.' From inside, even moored submarines give a powerful impression of being on the seabed, an effect only partly to do with the way daylight is rationed by a tiny hatch or two. It is one thing to go down into the depths in a bathysphere of one's own free will, like William Beebe, but surely quite another to go into combat in that blind tube crammed with men and machinery. There was in the *Bowfin* an air of menace greater than could be explained merely by the ever-present pressure of the sea beyond the curving steel sides. Other men's fear as well as extremes of discomfort, perhaps. In the engine room with the four huge diesels going for surface running it must have been hot even with the hatches open. When the submarine dived the engines were stopped and she went down under electrical power. In the already hot, confined space the four diesels went on giving up their heat. The engineers were naked but for shorts. Dehydration was a serious problem; men passed out.

The *Bowfin* produced a quite different effect from that of the *Arizona*. Although both ships had been in combat some half a century ago, the *Arizona* had felt as if she belonged even further back, to another epoch. In the museum were old photos of cocky *Bowfin* ratings lolling and smoking in port, draped around the very machine-gun up on deck against which I had just been leaning. Young in their 1940s hairstyles, they had none of the remoteness attending the marble names fixed above several tons of bones out in the harbour. For the real subject of Pearl Harbor is time, no matter how well it has been displaced on to trenchant exhibits. This at once became clear in the matter of gunsights. It was possible not only on *Bowfin* but also in the shore display to look through various gunsights, ranging devices, binoculars, periscopes and so on. Unlike much else these had not been maintained in working order. Since being given over to the public their focusing and other adjustments had become frozen or disconnected. At most there was a speckly view in one eye. Looking through a periscope towards the distant *Arizona* Memorial across the harbour I half expected to see a grainy, black-and-white clip from the film we had just been shown, as in the M. R. James ghost story about the pair of binoculars through which one could see only violent and desolating scenes from the past.[1] Indeed, I would swear I saw no colour through any of those eyepieces.

So the *Arizona* lies out there for all foreseeable time on the far side of a strip of water which is really too thin as insulation. Crossed as it is by a precise succession of launches bearing (among others) parties of faintly triumphant Japanese pretending to be from Taiwan, it scarcely cordons off the present. It all feels far too close, that nearby coastal strip, the freeways, used-car lots, dune buggies and 'All the Ribs You Can Eat for $5' joints. That, roaring by in its oblivion, does not feel like a ransomed world but one which has no use for the past in any other form than in national shrines.

It was at Pearl Harbor I first appreciated how, once it swallowed something, the sea washes it over less with water than with time, so whatever it engulfs becomes ancient almost immediately. It has something to do with being shut off from the continuity of vision, but in a way which is more powerful than mere burial on land. It felt as though the eight hundred skeletons contained by the still-leaking hulk of the *Arizona* had been borne back and out of history until they and the *Titanic*'s victims and those of the *Mary Rose* or any Phoenician galley were coeval.

---

[1] M. R. James, 'A View From a Hill.' The field-glasses were filled with black ichor distilled from a corpse, and to look through them was 'seeing through a dead man's eyes'.

With its polished metal and ruthlessly closing watertight doors, the *Bowfin* pulled me back to an event in childhood. When I was between eight and ten years old there was a terrifying disaster involving a submarine. That is, what I remember is my own version in which I imagined what it was like to be a sailor trapped on the seabed in a metal coffin, eyes raised in silence to the curved ceiling in hourly expectation of the first sounds of rescue. I could not remember the submarine's name, where it went down, or even the year.

What I am sure I recalled were the solemn tones of the BBC's news bulletins: the grave, Home Service accents coming through the varnished wooden slats of our old EkCo wireless. Why the submarine had failed to surface was unclear. It was one of 'ours,' in home waters, lying on the bottom intact but unable to come up and breathe. Memory has stretched the whole affair over many days, during which I invented everything we were not told. I had the impression of maybe one hundred ratings being informed by a level-voiced captain that if everyone kept as still as possible, breathing as slowly and shallowly as they could while all unnecessary heating and power were turned off, the oxygen could be eked out for maybe four or five days, even a week. Certainly long enough for rescue to arrive. Were not their comrades in the Royal Navy (up there in the sunlight) the most intrepid in the world? With the most advanced rescue techniques? Sit tight, the captain said; help is on its way. And—yes—pray, of course. More things are wrought by prayer / Than this world dreams of . . .

So we huddled over the wireless and kept pace with the rescue attempt as it grew ever more protracted and the BBC's voice ever more solemn. There was, we were told, a plentiful supply of 'oxygen candles' on board. These could be used at intervals to release fresh quantities of the life-sustaining gas. What these candles really gave off was *time*, of course: the extra minutes and hours fuming up as the men lay side by side in their bunks in the dim glow of emergency lighting, their whole world canted at an angle and with drops of condensation falling from the ceiling. Only a couple of inches of steel held back the press of black water outside. Some wrote letters home, others played quiet games of chess. Everyone was in good spirits, patient. No one wanted to be the first to say that although their prison was achingly cold, it was also becoming unbearably stuffy.

Until finally, after many days, we up there in the bright sunshine were told there was now no further hope for the crew of the lost submarine. The searchers knew where she lay but she was too deep. Or she was at the wrong angle. Or the special cutting equipment was still on its way from Rosyth. And somewhere hidden away in the black and icy depths the captain, his breathing now painfully laboured but his voice still level, would hand round to every man on board a little black capsule. The bulletins stopped abruptly and other news displaced the lost submarine. Ignored, it now lay in silence except for an occasional creaking as it stirred gently to a deep current. After another few days (but how long? Weeks, even?) and in the absence of a living hand to turn it off, the emergency lighting faded to a red glow and finally winked out. Only then did utter darkness cover the dead crew lying in orderly fashion in their bunks, with here or there an outflung arm or scattered chess set to betray a final struggle.

This was how, as a haunted pre-adolescent, I had imagined the drama to which the whole nation was made privy. Even now, some fifty years later, I can recapture an elusive wraith of the original terror and sadness I carried about at the time, and visiting the *Arizona* and the *Bowfin* jolted my memory still further. An atmosphere at once solemn and filmic inhabits one's contemplations of all sunken tombs, airless but watertight as they might variously be.

I also remember wanting to know the real details. What

happened to the bodies? In the absence of oxygen and in the near-zero temperature, how much would they decay? Would the submarine eventually fill with water from the combined seepages of hatchways and torpedo tubes and sprung plates? Before that would the batteries split, leaking acid to react with whatever seawater had pooled in the bilges and release chlorine? And would that in turn arrest any further bacterial activity, just as in tiny concentrations it could sterilise whole swimming pools?

Or maybe they had finally salvaged the submarine so that one morning it resurfaced, streaked with rust and shaggy with weed? Perhaps they had been advised to use caution as they released the steel dogs securing the main hatch and, as the last one was thrown, the heavy slab was hurled back on its hinges while a roar of putrid gases blew a column of rotting papers, naval caps and pocket chess sets high into the air. And once this dreadful pressure cooker had been opened, the first brave men wearing breathing apparatus and carrying flashlights would descend . . . Was that how it had been?

So many and efficient are the ways of deferring or obliterating curiosity in adulthood that it was not until visiting Pearl Harbor that I realised I still really wanted to know the answers. The nameless demanded to be named. I decided to track down this doomed submarine, to discover how much I and my friends had embroidered. For instance, it was inconceivable that the BBC would have described a submarine commander having a supply of suicide pills he could dole out when he judged a crisis hopeless enough, as if his men had been spies. After some research I narrowed down a handful of possibilities to a sinking which fitted all the criteria. In April 1951, HMS *Affray* went down in nearly three hundred feet of water off the Isle of Wight. Aboard her were seventy-five ratings and officers. She also, according to a contemporary newspaper report, carried 'a large quantity of oxygen candles'.

The *Affray* had sailed from Portsmouth on the evening of 16 April on a training exercise, part of which was to involve putting four Royal Marine commandos ashore on the Cornish coast. She was last heard of at 21.16 that evening, diving south of the Isle of Wight. Her commander, Lt. Com. John Blackburn, had been ordered to report daily between 0800 and 0900 hours. At 10 AM on the 17th, having heard nothing, the radio room at Fort Blockhouse, Portsmouth, alerted the authorities with the executive 'Subsunk' code. An hour later a search was under way which over the next two days would involve Royal Navy, US Navy, Belgian and French craft as well as the RAF. It was thought there was enough oxygen on board *Affray* to support the crew for three days, barring damage to the system. There was a suit of Davis escape gear for each man which included breathing apparatus and immersion suits.

That night other submarines reported asdic contacts and Admiral Sir Arthur John Power, Commander-in-Chief Portsmouth, announced '*Affray* has been located on the bottom thirty-five miles south-west of St. Catherine's Point in just over thirty fathoms of water.' At daybreak an aircraft dropped small explosive charges to tell the submarine's crew she had been found and that ships were standing by to pick up anyone who came to the surface. Nobody came. Several of the searching submarines reported hearing faint, distorted signals and sounds which might have been made by someone tapping on a hull, but nobody managed to get a reliable bearing. In the afternoon the asdic room of HMS *Ambush* picked up the code letters which meant 'We are trapped on the bottom.' By this time thirty-four ships were taking part in a search which was becoming desperate.

Next morning hopes were raised again when an RAF Coastal Command aircraft spotted oil and dropped a smoke canister which was mistaken by another plane for a marker buoy from *Affray*. That evening, sixty-nine hours after she had dived, the submarine was officially given up for lost.

Because the *Affray* was only one of sixteen 'A'-class submarines it was vital that the reason for her sinking be ascertained. For the next eight weeks the search continued under Captain W. O. Shelford. By early May she still had not been found, and Shelford was reluctantly driven to take notice of the large number of letters and phone calls being received at Portsmouth from members of the public who claimed to know where she was. He plotted these alleged positions on a chart and found to his surprise that they mostly fell within a small area outside the main search zone.[1] Shelford told Admiral Power, somewhat hesitantly, and a ship was dispatched to investigate. When it arrived at the location it at once obtained a strong asdic echo; yet as it turned out it was not the *Affray* at all, nor any other wreck or rock.

Probably every such sinking generates its own aura which profoundly affects not merely the public at large but those involved in the search. There is an account of a strange experience which the wife of a Rear Admiral at Portsmouth had on the night of 17 April, the first evening of the search for *Affray*.

Quite suddenly, I realised that I was not alone in my room and in the half-light I recognised my visitor. He had been serving as an engineer officer in my husband's ship, a cruiser, at a time when my husband was an Engineer-Commander, and we had often entertained him in our Channel Islands home.

He approached me and stood still and silent; I was astonished to see him dressed in normal submariner's uniform although I did not recognise this fact until I described his clothing to my husband later. Then he spoke quite clearly and said: 'Tell your husband we

[1] For this and many other details of the sinking of *Affray* I am indebted to the Captain's own account in W. O. Shelford, *'Subsunk'* (London, 1960).

are at the north end of the Hurd Deep, nearly seventy miles from the lighthouse at St.. Catherine's Point. It happened very suddenly and none of us expected it.' After that the speaker vanished.[1]

The lady had promptly phoned her husband, who said he had no idea this officer had even transferred to the submarine service, still less that he was aboard HMS *Affray*. Since the Hurd Deep was well outside the main search area, ships could hardly be diverted on the basis of a ghost story. That, of course, was while there was a chance the crew might still be alive. It was to be some weeks before Captain Shelford gave enough credence to the other seers and clairvoyants to tell his superior.

Affray was eventually found by HMS *Reclaim*, using an under water TV camera, a new technology's first major success. She was lying in forty-three fathoms of water on the edge of the Hurd Deep with a slight list to port. It was sixty-seven miles from St. Catherine's lighthouse. At first sight she appeared undamaged. All hatches were shut and none of her indicator buoys had been released. Her hydroplanes were set to rise. Then serious damage was found to her snort tube. This was a hollow mast, thirty-five feet tall, through which the diesel engines could breathe, enabling the submarine to run on her surface engines while shallowly submerged. The cameras found it was almost completely snapped off at the base and the hull valve inside was open. The mast was winched up and examined. It was thought at first there might have been a collision, but it was soon found the fracture had been caused by the design not having been strong enough, combined with faulty material. Probably her commander had radioed his last message, dived, and the snort tube fractured on the way down. Even at a depth of only 40 feet, water would have poured through the open valve at a rate of three-quarters of a ton per

[1] Quoted in Warren Armstrong, *Sea Phantoms* (London, 1956) and Edwyn Gray, *Few Survived* (London, 1986).

166 - JAMES HAMILTON-PATERSON

second. It would have been impossible to have closed the valve against such a flow even if somebody had been standing by it. The water would at once have flooded the engine room and caused electrical short circuits followed by explosions, fire and the release of noxious fumes.

So it seems after all that the *Affray* was overwhelmed quickly and without warning, and that by the time the alarm was raised her crew had already been dead several hours. The 'signals' heard during the search, the code word, the tapping, all were imagination on the part of anxious young men with headphones clamped to their ears in submarines identical to the one that had vanished. The powerful asdic return from the clairvoyants' recommended 'position' was almost certainly from the DSL, or Deep Scattering Layer.[1] As for the Rear Admiral's wife, her visitant had been completely accurate.

It would just have been possible to salvage *Affray*, given her depth, the currents at that point and the technical capabilities of the day, but it would have been expensive and dangerous. 'Since the cause of the disaster had been established little was to be gained from such an operation, her scrap value being no more than about £5,000. Salvage was abandoned.'[2] In 1990 her wreck was resurveyed. An officer in the Wrecks Department at Taunton Hydrographic Office added that he himself had been aboard one of the submarines searching for *Affray* and the story of messages being picked up by asdic was 'quite without foundation.' He said the *Affray* had flooded instantly. 'It was chaos out there—messages whizzing about—and no doubt some people imagined what they most wanted to hear. It was a very emotional business, of course.' As for the 'suicide pills': 'Absolute poppycock. Never been such a thing. Pure bosh.'

Concerning the bodies themselves, my schoolchild's fantasy had presupposed a watertight submarine. It is possible to form

[1] See Chapter 5, p. 189.
[2] *Brassey's Naval Annual* (1952).

a few tentative theories as to what might happen in the circumstance of seventy-five people dying of oxygen starvation. Much would depend on things such as temperature and the fungi in the remaining atmosphere, though it seems likely the air scrubbers would have removed most floating yeasts. From that point of view a submarine is probably a comparatively sterile place. In any case the hypoxia and high carbon-dioxide levels would initially accelerate decomposition due to venous congestion. This would be followed by a slow process of mummification during which the skin hardens and the gaseous cavern of the stomach contracts. It is likely that adipocere would form, especially in dependent limbs. This is the condition when fat changes to an off-white, waxlike substance smelling slightly rancid or musty. Adipocere itself being preservative, this would prevent further decay in affected portions. Provided a sunken submarine remained watertight, a trip down her interior with a flashlight would probably reveal a majority of mummifying bodies, some partly decomposed, a few even skeletonised entirely. Here and there where a head had been subject to adipocere a face might be seen whose features had scarcely changed.

In the case of sudden flooding, as in the *Affray*, things would be entirely different. The first, powerful inrush of water would have caused extensive injuries, including rupture of the eardrums by the rapidly increasing pressure. Since any opening, even one only ten inches across like a snort valve, would give access to marine animals the stripping of the bodies would begin within hours. In the sea it is generally the lips, eyes and fingers which go first, being most easily seized by creatures with small mouths or pincers. Cod are especially voracious and as soon as the seawater softens the flesh the remainder will be torn off quite rapidly.

Except in terms of size, the *Arizona* is no more of a tomb than the *Affray*, but its status is quite different. It is a national shrine because it fell victim at a turning point in US history, on 'a date which will live in infamy,' in President Roosevelt's words. Part of her power as a symbol comes from being visible but inaccessible. One can touch a mausoleum; a relative might put flowers on an actual grave in Arlington National Cemetery and pass a musing hand over the carved name. But no relatives may touch the *Arizona*. Not even the memorial from which they gaze down touches her. Everything about the ship has passed out of the realm of the personal. A further part of her power derives from being monumentally a heap of junk. Most war debris is cleared away, especially if it is blocking a harbour. To have left this hulk in the face of expediency or even aesthetics (for, shorn of its symbolic value, there is nothing very beautiful about bits of rusty steel poking up above the water) is a powerfully contrary gesture. It is a solemn act, going against every urge to tidy away, clear up or edit the past. This being the case, it is hard to imagine what might happen to any blithe infidel who took it into his head to don scuba gear and loot the *Arizona*.

The *Affray* is in a different category altogether. She was a peacetime casualty, though still a military craft. She symbolises little beyond the misfortunes of happenstance. Most of her interest lies in a collective, brief but intense involvement with the race to save her when, in fact, she was already dead. She is a tomb to seventy-five and cannot be tampered with, though without the same taboo as the *Arizona*. In fact, the whole issue of when undersea tombs can and cannot be touched, and to what extent, is complex. In March 1991 three men with a good deal of equipment dived on the wreck of another British submarine, the *E 49*. This had gone down with all hands in 1917 off the Shetland Islands and was first found by local divers in 1988. The men looted one and a half tons of assorted bronze fittings worth $2,300 and later, before a court in Lerwick, admitted to 'theft by finding.' It was stated in court that the wreck belonged to the Ministry of Defence and the men pleaded guilty while claiming they did not know the items were part of a submarine. Two of them were lightly fined and the third was 'admonished.' At no point in the newspaper reports was there any mention of the submarine's being a tomb or a war grave.

On the other hand, the *Titanic*, which sank five years before the *E 49*, at first acquired untouchable status though only once the technological means for touching her had been perfected. But before long a few artefacts were dredged up and shown on French television. Dr. Robert Ballard, who led the 1985 expedition to film the wreck, has always made a point that nothing should be removed from it, to the extent of throwing back a piece of cable snagged and brought to the surface by an unmanned reconnaissance sled. (He has, however, added to the wreck: a bronze plaque from the *Titanic* Historical Society commemorating the ship's 1,522 dead.) It remains to be seen whether others will be as scrupulous. The ship's best defence is her depth, the secrecy of her exact coordinates and the huge expense of all exploration and salvage technology.[1]

The question stands: when does a wreck pass into the gaze of archaeology enough for the numbers of dead on board to be irrelevant to investigation and salvage? Any famous wreck belonging to this century might be thought safe, on the grounds that victims' relatives might object or sue. Yet that cannot be a hard and fast rule since several modern wrecks

---

[1] Since I wrote this paragraph Robert Ballard's expressed hope that 'world opinion' would respect this noble graveyard and foil any plans to exploit the wreck has been shown as vain. *Titanic* has since been thoroughly disturbed by wholesale salvage, scientific sampling and tourist trips. Recently, the methodical looting has been further legitimised since the timely announcement that the wreck is fast decaying. Salvage (especially of the ship's safe, when it is found) can now be presented as a race against time. This gives the registered salvors, RMS Titanic, carte blanche to strip it in the name of 'archaeology,' even though the ship is not yet a century old and most of the objects retrieved so far are the sort of junk we threw out of our grandparents' attic when they died.

have been quite officially looted if they were carrying sufficiently valuable cargoes. There must be some unwritten algorithm which balances the number of bodies against bullion or lost art or artefacts, which only takes shape in words when it is already too late.

Military wrecks fall into a separate category since in Britain, at any rate, they belong to the Ministry of Defence. This is not a claim likely to be relinquished, moreover, since in certain respects the scrap value of wrecks increases as time goes by. Some fifty years on, the *Affray* would assuredly be worth more than $7,500. World War I craft have an additional value in that their steel dates from the pre-atomic age and contains no radioactive isotopes such as of caesium. This is known as 'pre-atomic' or 'aged' metal and came into demand as a direct result of the numerous H-bomb tests in the 1960s. These caused atmospheric pollution in the form of short-lived nucleides which have contaminated every forging or casting made anywhere in the world since then. When scientists wanted to be able to measure very low levels of radiation in samples they were testing there seemed to be no way of making a screen to exclude normal background radiation since the steel of the screen was itself contaminated. The solution was to salvage the gun barrels of pre-atomic warships. Their advantage was that they did not require re-forging. The barrels were simply cut into lengths and the scientists could lower their samples into these massive, inert chambers. For many years demand for aged metal has been satisfied by the wrecks of the German fleet scuppered at Scapa Flow, a rich source of uncontaminated steel as well as of copper, brass and nickel-alloy armour plate, which has its own brokers. However, the need for pre-atomic metal has already begun to lessen with the decline of atmospheric nuclear testing and the decay of short-lived isotopes.

There is obviously nothing new in the idea of bringing up valuables from wrecks. Livy mentions it, and in Rhodes there was a law of salvage which apportioned reward according to the depth from which the treasure was retrieved. In water twelve feet deep, a diver received a third of its value, at twice that depth a half. Great fortunes were made by men working with very primitive apparatus. In 1667 William Phipps used wooden diving bells to work in water up to sixty feet deep while recovering $300,000 from a galleon sunk off the coast of Spain. For this he received one-tenth, while $130,000 went to the Duke of Albemarle and the remainder was distributed to the enterprise's other subscribers. There seems to have been no claim made by the Spanish Ministry of Defence, nor any particular respect paid to the bones of the crew.

It will be seen how readily, at certain points, three of the categories of wreck—tombs, time capsules and gold mines—elide into each other. In time almost any wreck becomes valuable. Sheer treasure will generally outweigh scruples. The perfect example of a wreck too recent to be of any real archaeological interest and too old to be thought of as anyone's tomb is that of the SS *Central America*. This sank two hundred miles off the coast of South Carolina in a hurricane in 1857. Of its 600 passengers, 425 went down with her; but they did not prevent the cargo of Californian gold from being raised recently. Probably the single richest treasure ever found, it was worth a billion dollars. 'Found' in cases like this does not mean the serendipity of just happening to stumble on a fortune. More searching can take place in libraries and record offices than on location.

It may well be that one of the proprieties of salvage hinges on something as banal as whether or not there are any skeletons left. The action of the sea on human bones is referred to in the next chapter; in brief, much depends on salinity and especially on pressure—in other words, depth. Unloading a seemingly deserted hulk would be quite different from looting beneath the knowing, sardonic gaze of skulls. Like the *Central*

America, the *Titanic* does not yet have for us the archaeological interest which made the recovery from the Solent of the *Mary Rose* worth undertaking when it was quite clear that, as a warship, she would not have been carrying treasure. Besides, it had been declared in advance that any artefacts found would go into museums and not sale-rooms. But before long, *Titanic* will be comprehensively looted. It has a mystique about it like that surrounding the tombs of ancient Egypt. What fascinates is not the possibility of finding the odd tiara so much as the retrieval of domestic objects impregnated with this mystique. To be able to eat off one of her dinner plates decorated with the shipping line's emblem of a white star inside a red burgee would be strange until it became a commonplace. Around a century on the seabed endows an ordinary utilitarian object with a reality slightly at one remove. *It ought not to be here, but it is.* Yet part of it is still in the timeless realm into which it disappeared. Tutankhamun's trumpets have been played after 3,300 years of silence, and if their haunting sound comes at us from another world entirely, so would that of—say—a trumpet belonging to one of the *Titanic*'s band which had last played 'Abide with Me' on her steeply canting deck.

Lying off a Philippine province there is a wrecked British cargo ship of the mid-nineteenth century from which storms occasionally wash thick English crockery on to the beach. There is probably nothing of much value on board since the vessel was merely supplying the outposts of sugar companies on Negros. It is odd to see old stone beer bottles so far from home and used as oil lamps in the local huts, as it is to sip palm wine out of china mugs with 'Staffs' stamped on the bottom beneath the glaze. The sea seldom gives back much of what it takes and when it does, in random and haphazard fashion, the effect can be striking.

Generally, the sea hides and dissolves, and in translating objects from the upper to the lower world, it obscures with 'an immensity that receives no impress, preserves no memories, and keeps no reckoning of lives.'[1] It is into this immensity that aircraft as well as ships disappear. Somewhere under the waves are the last fragments of Amelia Earhart, Amy Johnson, Glenn Miller, Antoine de Saint-Exupéry and countless others; to say nothing of the redoubtable Duchess of Bedford who was last seen heading out alone over the Wash in her De Havilland Moth in 1937, aged eighty-one.

As for time bombs, these are usually objects which people prefer not to think about too closely. They include literal ones, like the munitions ship *Richard Montgomery* which sank late in World War II in the mouth of the Medway with a full cargo of bombs, shells and high explosives. It is still there, visible and clearly buoyed off Sheerness. It has been estimated that if it ever does explode it could level the town, and local MPs periodically raise the matter in parliament. Consensus expert opinion remains that the explosives have reached such an unstable condition it is as well to leave the ship alone. Time bombs of a different order are the drums of toxic chemicals such as dioxin which are frequently washed off the decks of cargo vessels during storms or dumped as part of an effort at 'waste disposal.' Radioactive waste is similarly still being dumped at sea in reinforced concrete containers. Oil tankers also go down from time to time without all their tanks rupturing, so that somewhere, conveniently hidden, lie millions of tons of crude oil waiting to escape. Entropy being what it is, and oil being lighter than water, sooner or later they will.

Finally, there is perhaps a fifth category of wreck, one which combines the characteristics of tomb and time bomb in that for one reason or another people do not wish to investigate further. On the night of 20 December 1987 there was a collision between two ships in the Tablas Strait off Mindoro in the

---

[1] Joseph Conrad, *The Shadow-Line* (1916).

Philippines. The larger of the two vessels was the M/V *Doña Paz*, a passenger ship belonging to Sulpicio Lines. She was grossly overladen, an everyday occurrence in itself but made worse than usual by the approaching holiday. Somewhere off Dumali Point she collided with a lightless and rusty tanker, the M/T *Vector*, owned by Caltex. There was a huge explosion and a fireball which was seen by fishermen forty miles away.

From the *Doña Paz*, twenty-six survived: twenty-five men and a girl of eighteen. From the *Vector*, two. To this day nobody knows exactly how many died. Three thousand is the official estimate, but it is certain to have been many more. The passenger ship's manifest showed only the legal maximum, of course, but that will not have been a true count even of those families listed since children under the age of ten are never included as they travel free. Possibly the dissolving bones of 3,500 people lie in over 1,500 feet of water in the Tablas Strait, but nobody wants to be reminded in too much detail of the world's worst-ever peacetime disaster at sea.[1] At the time, world leaders, including the Pope, expressed 'anguish.' President Aquino ordered an immediate 'all-out probe.' Not until November the following year—and only after another accident involving one of its vessels (see Chapter 7) – were the shipping company's operations briefly suspended. The captain of the *Doña Paz*, who did not survive, was thought to have been drunk and playing mah-jong at the time of the collision. It was said that the *Vector*'s bridge was completely empty, and the charge filed against Caltex was for carrying 'a highly dangerous mix of cargo in a grossly inadequate and unseaworthy vessel'.[2] Meanwhile, the victims' relatives formed an association to press for proper compensation. Nearly eighteen months

[1] Some twenty years after the collision, a death toll of 5,000 is now generally accepted. Some maintain it was still higher, including the parents of the two boys who are this book's co-dedicatees.

[2] *Manila Bulletin*, 20 December 1987, *et seq., passim*.

later, in May 1989, Sulpicio Lines claimed '86% of the passengers aboard the *Doña Paz* have been paid for at ₱ 30,000 per victim' (about $900). That is, eighty-six per cent of those listed on the manifest. The company admitted no liability for the passengers it had carried illegally and who could not now be identified. Also, of course, the affair had incited hundreds of completely fictitious claims. One way and another, it was convenient that the sea should efface the evidence and close for ever over the President's 'all-out probe'.

II

A stitch through the nostrils

The common fantasy of the sea's withdrawing or vanishing to reveal the naked ocean floor shows its involvement with loss. From this derive an especial melancholy and a power to haunt. Among the sea's attributes are a capacity to conceal, the ability to stand for time, and the quality of erasure.

The ocean's capacity to conceal does not extend only to things one might lose overboard, or to the people, ships and cargoes it has always swallowed up. Its deeps may hide the unknown and the monstrous but they can also veil from sight places which have never been lost but for which some are always searching: Atlantis and those far-off worlds whose presence is less a cartographer's fiction than a fact of desire. This is not to disparage the imaginative space they occupy but to wonder if the future identification of a particular lump of seabed with the fabled land, the Happy Isles or Mayda itself might, after the initial euphoria of vindication, not please its supporters as much as they thought and even mean their having to find something else on which to fix their search.

There is a need for the myth of the lost land which so often has utopian or golden significance. Interest in a submarine Atlantis grew as increasing exploration and travel removed areas of dry land which might plausibly still conceal Golden Khersonese, Ophir, El Dorado and the sundry lost cities of higher civilisations. (At the turn of the twentieth century Krupps of Essen spent half a million dollars looking for El

Dorado in the Mato Grosso, while in 1925 Lt. Col. Percy Fawcett was murdered in the Brazilian jungle by Calapalo Indians while looking for the ruins of Atlantis or one of its sister cities.[1] As for Conan Doyle's *The Lost World* (1912), it is significant that he visualised it as an island, a raised plateau surrounded by a great escarpment of practically unscalable cliffs.

The loss of the *Titanic* was not a myth, yet both she and the search for her have taken on something of a mythic status, combining semi-archaeology with the qualities of a quest. Since this is a secular age, sacred relics will no longer do as quest objects (the demotion of the Shroud of Turin from holy trophy to medieval forgery ought to have dealt the final blow to the sacred-object industry). Things swallowed by the sea will do excellently in their place, however, especially if there are TV rights to their finding. Dr. Robert Ballard's search for the *Titanic* was on his own admission obsessive, and such things arouse wide interest. 'My lifelong dream was to find this great ship, and during the past thirteen years the quest for her had dominated my life.'[2] It had also cost huge sums of money. Yet the ideal thing about this quest was that it could be rationalised by turning it into a research project. In the 1970s, the Woods Hole Oceanographical Institution decided to increase the depth range of its deep-sea submersible, *Alvin*, from 6,000 to 13,000 feet. Since 13,000 feet was roughly the depth at which *Titanic* was supposed to lie, searching for it would be the perfect way to test the improved submersible and a new generation of robotic vehicles which could be deployed from *Alvin* on cables and guided by remote control. These in turn

[1] For a scholarly treatment see L. Sprague de Camp, *Lost Continents: The Atlantis Theme in History, Science and Literature* (New York, 1970) and Paul Jordan, *The Atlantis Syndrome* (Sutton, 2001).

[2] Robert D. Ballard, 'A Long Last Look at *Titanic*,' *National Geographic*, December 1986.

would be prototypes of the entirely free-diving robotic sleds which are intended eventually to replace manned submersibles altogether. The enterprise was further legitimated by first being put to military use. The Woods Hole research vessel *Knorr*, needing to try out the *Argo/Jason* equipment Ballard had helped develop, made a practice run over the wreck of the US nuclear submarine *Scorpion* which was lost with all hands in 1968. The wreck was comprehensively filmed though the pictures are still classified and have not been released. Thus exonerated, Ballard could turn his attention to the *Titanic* and later that same year, 1985, did indeed find her.

It is a happy man who can spend other people's money and indulge his own ingenuity to fulfil a lifetime's ambition. The triumph of Heinrich Schliemann when he discovered Troy stood for Freud as the perfect image of happiness, but it also begged questions about what exactly was being satisfied. The problem of Troy was that it had disappeared to the extent that people wondered if it were mythical, or maybe a composite like Homer himself. To have proved it existed and to have stood in its ruins must have been even more exciting than finding Tutankhamun's tomb was for Howard Carter. Troy had been legendary for 3,000 years whereas the young king was a footnote in dynastic history. Carter had only dangled his name in front of his sponsor, Lord Carnarvon, to induce him to let him have a last dig in the Valley of the Kings, which by then had been turned practically upside down in the search for dead pharaohs.

Dr. Ballard's quest was like neither of these, precisely, in that it was not an act of archaeology. Everybody knew the *Titanic*'s fate; it was no legend. Nor was there any mystery about how she had sunk, as in the case of *Affray* and *Scorpion*. Bland reasonableness ('It's down there somewhere and I'm going to find it') is always unsatisfactory as an explanation for a life's dream and thirteen years of searching. When in such cases the search is described as being more important than finding the object one is entitled to ask what is really being looked for. Such avidness is normally reserved for things of great personal significance that one has lost oneself, and soon gives way to resignation. What private thing, in short, did the *Titanic* stand for? The question must be left unanswered, but it should be asked. What can be said of such ventures is that they seldom stop there.

Dr. Ballard did indeed go on to find the *Bismarck* and, no doubt, much else besides. In the long sequence of searchings and findings, searchings and findings such men undertake, success is a temporary setback, a resting place on a much larger and grander journey to find the one thing that will satisfy a loss which can never be specified. The sea is the perfect place for it since whatever it hides beneath its dark leagues of surrogate tears it makes timeless. One might announce one was looking for a lost submarine but a thick, wounded shadow dimly glimpsed at the edge of a monitor screen would seem a thing immeasurably ancient in its melancholy, weeping 'rustsicles' from its iron plates, and the thrill of discovery would always carry with it the minor guilt of intrusion.[1] It would also be imbued with the knowledge that what one finds never fits the cavity which the search hollows out. 'In a way,' said Dr. Ballard of the *Titanic*, 'I am sad we found her.' Such hulks will always feel privately significant to their finder. At first they may appear to be playing a game with him, being deliberately, almost flirtatiously, elusive. Later they become in their stately woe repositories for his own unassuageable loss. So it was that Dr. Ballard identified with the object of his quest to the extent

---

[1] This observation proved oddly anticipatory. Four years after I wrote it I found myself 5,000 metres down in a Russian submersible, searching the Atlantic seabed for *I-52*, a Japanese World War II submarine. For me, the search was flavoured with exactly this mixture of longing and dread. See *Three Miles Down* (1998).

of expressing anger at the teredo worms which had devoured her woodwork. 'After years of gluttony the creatures starved and dropped dead at the table. I have no sympathy for them . . .'[1]

In the circumstances it is perhaps not surprising that so much emphasis should be laid on *not touching*. This perpetually lost object cannot be touched because at that instant it will turn into something else: an ordinary ship, an ordinary battleship, which sank, which has a salvage value, which will attract looters. Not to touch leaves it exclusive, an object of vision, in some sense still not wholly found, while heightened as the public myth. The underwater camera records the details, the mystery remains intact. How could one not sympathise with a man who dreads to think his adroitly publicised private quest might merely stimulate greed rather than a proper solemn wonder? It would be as if Sir Percival were to learn that his companions on the Quest only wanted to find the Holy Grail so they could melt it down.[2]

Given the sea's power to swallow and erase it is not surprising that it should have been seen as the lair of malignant spirits. The British sailor's personification of these was Davy Jones, to whose locker the drowned were consigned. He is thought by some to derive from 'the Devil Jonah.' Jonah was thrown overboard as a sacrifice by his shipmates to pacify the storm for

[1] Ballard, 'A Long Last Look at *Titanic*'.
[2] In the epigraph Robert Musil chose for his novel *Young Törless* (1906), Maeterlinck wrote:
> In some strange way we devalue things as soon as we give utterance to them. We believe we have dived to the uttermost depths of the abyss, and yet when we return to the surface the drop of water on our pallid fingertips no longer resembles the sea from which it came. We think we have discovered a hoard of wonderful treasure-trove, yet when we emerge again into the light of day we see that all we have brought back with us is false stones and chips of glass. But for all this, the treasure goes on glimmering in the darkness, unchanged.

which they held him responsible since he had angered God by disobeying orders to go to Nineveh and remonstrate about its sinfulness. Jonah, preferring a quiet life, had slipped away to Joppa instead, where he found a boat bound for Tarshish at entirely the other end of the Mediterranean. His locker turned out to be the stomach of a whale. The idea of sacrificing somebody to a storm god was no doubt ancient. What is curious is that the practice has lingered for so long, commuted to the passive form of refusing to save the drowning or to touch the drowned. Bryce, the pedlar in Sir Walter Scott's *The Pirate*, observed 'To fling a drowning man a plank may be the part of a Christian; but I say, keep hands off him, if ye wad live and thrive free frae his danger.' This is not quite the spirit of Grace Darling.[1] Several widely separated cultures have been recorded as sharing the refusal of certain Scots villagers to pick up a drowned body for fear of meeting the same fate. A description of such behaviour can be found in a nineteenth-century account of a ramble in Austria. It is true that the incident concerns the River Danube rather than the sea, but the principle is undoubtedly the same:

> In Upper Austria I witnessed an unfortunate crew member knocked overboard by the tow-rope. He fell into the waves and eventually drowned, but not without a long struggle. During this the rest of the crew, or his shipmates, made not the slightest attempt to rescue him. Showing every sign of not wishing to become better acquainted with the nether regions of the Danube,

[1] Grace Darling (1815–1842) was the daughter of the lighthouse keeper on the Farne Islands, off the Northumberland coast. On 7 September 1838 a steamboat was wrecked in a storm and Grace and her father rowed out and rescued survivors from a rock. Practically overnight she became a heroine; trust funds and awards were showered on her; a circus made her an offer she had no difficulty in refusing. 'Applications for locks of hair came in till Grace was in danger of baldness' (*DNB*, 1975). She died of tuberculosis four years later, unspoilt by fame and unsullied by marriage.

and probably knowing nothing of the local Lorelei, he fought the waves' clutch to his last gasp. Meanwhile, the crew leaned unmoved over the rail and called in monotonous chorus: 'Give up, Jim; it's God's will.' This went on until their comrade disappeared.[1]

The author goes on to explain that despite his jocular reference to the Lorelei, local superstition was not concerned with legends of water nymphs beckoning sailors to join them in ravishing underwater kingdoms. Indeed, the boatmen seemed fatalistic rather than superstitious. They simply believed that the river required at least one human sacrifice a year and attributed this directly to an ordinance of heaven.

Such things testify to yet another unfixable boundary, that which the sea mediates between life and death, between air and water. It is not always clear at what point this is finally crossed. For some cultures and at some periods it seems that merely falling into water is enough to show a victim he has been 'earmarked,' that from then on he belongs elsewhere. In other places and at other times people are heroically snatched from beyond this flexible threshold and laboriously brought back to life. Timid swimmers may experience the boundary acutely, their upper halves in air and light, their lower in water and feeling the pull of black depths. It is a common enough description: 'I feel as though I am being dragged down,' as though a great magnet on the seabed were tugging at them, or as if they had entered the gravitational field of a private fate. Others, at home in the water, can hardly swim down far enough for their own delight.

Once, on an island in the South China Sea, I was taken to visit a *mangkukulam*, a sorcerer, who lived inland in an isolated hut by a stream. I was told he had captured a sea devil

---

[1] August Ellrich, *Genre-Bilder aus Oestreich* (Berlin, 1833), p. 12.

which was responsible for drowning several local fishermen. It was dead now, they said, but I was on no account to touch it because it still had some of its old power and anyone who touched it ran the risk of being dragged down. Only the sorcerer himself was safe because his power was greater and anyway he hardly ever went near the sea. This man, who turned out to be no more than thirty, produced a flattish bundle of purple cloth with a faintly ecclesiastical look to it, something like the material Filipinos use to make the stiff little copes and robes for their Santo Niño images. As he unwrapped it my companions edged backwards out of the hut into the daylight, peering through the open doorway. Just before lifting off the last fold of cloth the sorcerer muttered a little chant, then exposed a horror I was only half prepared for.

It was grotesque, about twenty inches long and not quite as wide, black, wizened, and with a terrible face. The mouth was a greedy smile, like a child's drawing or the segment of an orange, while the eyes slanted malevolently and shone dull red. It had four limbs like flippers, a miniature manatee's, slightly curled as if it had died trying to grasp something perhaps the size of a swimmer's leg. So imperceptibly can one take in, over the months, a small community's terms of living that it actually looked to me like a devil, just for an instant, before I saw it was the deformed body of a male stingray. The sorcerer had cut it, moulded it, *sculpted* it, setting the gill slits with glass beads and creating from the claspers pseudo-limbs so that the underside—which was on display—became a single nightmarish face. It was, in fact, an excellent example of a Jenny Haniver, a class of grotesques modelled from the bodies of real animals so as to resemble some legendary or imaginary creature. They were very popular in Europe in the sixteenth and seventeenth centuries and many a mermaid was formed (like that which relatively recently was still on display in a museum in Alexandria) from the doctored body of a dugong with long tresses of bleached horsehair and fake pudendum. It was the first I had seen and I was not surprised at the fear it aroused. Even the sorcerer himself seemed in awe of his own creation as he covered its potent eyes with cloth and wrapped it hurriedly up again.

It is not easy to say precisely what it is about the sea which so swiftly and powerfully puts things on a mortal footing. Sometimes drowning can seem the least of its lethalities. Even people who live by it and work on it all their lives will occasionally glance up and feel their thoughts and preoccupations momentarily effaced, swept by a bleak melancholy which they sense does something at once banal and radical, like resetting the poles of their existence. It is over in a flash; but the seething waters of the middle distance have something wordless and unpitying to say to the dullest of minds. It is as if the ocean, in certain lights and weathers, were the lair not of monsters or malevolent downward-dragging spirits, but of the fat blank which squats beneath all happiness, flicking out its tongue.

It can happen without warning. We might be travelling from one town to another without giving much thought to where we are, beyond being in a car or train. We look up and catch an unsettling glimpse of sea between hills. At once the day changes, becoming particular, much as it does when we hear of the death of a distant acquaintance. Suddenly we are made aware of a horizon, and over it has come a reminder of outlying importances which have always been there but have been temporarily forgotten. Merely knowing that the sea is near gives the landscape a different feel. For the rest of that day the edge of things nudges in towards us.

To most Europeans, at any rate, 'the seaside' is no mere littoral, a bald geographical margin where land happens to stop. It is too closely bound up with the past to be an indifferent location: with summer holidays, once-yearly pursuits, even erotic adventure. We delve in a cupboard in a winter month, looking for something, and come up instead with a pair of shoes from which sand trickles as from a snapped hourglass. At once we are flooded with memory. We almost hear surf break outside the window in waves which have crossed fields and city streets to find us. Summer's gear with its sad charge is an authentically modern artefact.

Driving to a holiday place on the coast from a town or airport creates an expectancy which is both fulfilled and intensified by seeing the straight line of ocean off at an angle between trees. *Someone lives here all the year round* used to be one's private thought, but only half envious. That straight line is always of another time, of another land where maybe once one had been happy. Else it stands simply for promise, that tireless harbinger of loss. Anyone who travels is reminded: over just such a horizon is . . . the land where lemons bloom, where corals lie, the El Dorado or Atlantis of the future.

So coastal towns have quite a different feel from those inland, poised for ever on the edge of onward travel or of turning back. They exude impermanence, as if everybody there were touched by this crucial indecision. Even their fabric is subject to it, as though the houses themselves knew that sooner or later they would find themselves eroded away or else stranded far inland. The old port of Dunwich, sometime capital and commercial centre of East Anglia, began to collapse into the sea in the middle of the eleventh century and now lies beneath the waves. And on the opposite side of the North Sea are Dutch fishing villages miles from the ocean, cut off from it by the building of the Great Dyke in 1932. (Tarshish, come to that, is today buried beneath the marshes inland from the mouth of the Guadalquivir River, north of Cadiz. In Jonah's day, and well before that, it had been an entrepôt on a bay which stretched inland almost as far as Seville.) The idea of no abiding city is nowhere more evident than on a coast, where over the centuries the sea writes and rewrites its own margins.

Maybe this is why in Britain so many old people migrate to the seaside to await death. In order that it should not be too obvious they are drawn to the impermanent to gaze at annihilation, the towns are often built to look deceptively solid and are sited in places renowned for their therapeutic qualities. Yet to retire to Bournemouth is to admit to being in transition. There they are, chairs pulled up in long lines on the promenade, passengers on a municipal voyage. Maybe for some it is a return to a deep, inchoate thing which has been slumbering inside them all their lives. It is death which stretches before them to the horizon, a great absolving sheet beneath which they will slip. What is there to see which makes them stare long hours? The constant flux of waters holds something that mesmerises *Homo*, though whether it speaks of human origins or of individual destiny is unclear. Moving water has in it a fascination both lulling and imperative. Maybe all continuous movement, whether flames in a grate, crowds in a street, trees in a wind or the flicker of a television screen can catch at the mind and set it into introspective motion. Of all such things, only the ocean never moves without an underlying gravity, even on waveless days of sparkle and dance. No weather is inappropriate for a burial at sea.

Death as a voyage is a common trope and the sea invites embarkation whether the dead are literally set adrift in a boat with a few possessions or sewn into a canvas shroud with the last stitch through the gristle between the nostrils and committed to the deep. Some, still on shore leave, may be seen drawn up on the council's benches along a pier. Before life goes out with the tide the waiting ranks of the elderly may yet be quite unmournful since their conscious intention in migrating to these, their last resorts, was anything but morbid. Rosily remembered childhood treats and holidays at Blackpool, Margate and Skegness awake hopes of dignified rejuvenation. By retiring to the edge of things a lifetime of unfulfilled summer wishes might be made good or truced. It is rationalised by talk of a milder climate and sanctioned by doctor's orders. Yet even the retired mind must know that most of the year is not summer. In the long winter months after dark, when invisible below the esplanade a black sea raps peremptory knuckles on the shingle, it is time at last to go.

Part of the pleasurable melancholy of beachcombing comes from speculating about where the objects came from, what they were, how long they took to arrive. Having been in the sea, jetsam, like wrecks, becomes pickled in agelessness. Even bright fragments of plastic give an impression that they might have been adrift for centuries or a week. A knowledge of winds and currents adds a further dimension of interest and I have spent hours like a maritime Holmes, pacing beaches and building up a mental map of how rubbish circulates in a complex archipelago. Once after three days of storm I walked a deserted coast and came upon the shelving mouth of a watercourse bringing floodwaters down from the hills in the far interior. The sea, still fretful, was tumbling in the surf the things it had thrown up together with what the river had brought down. Scattered around were sodden coconuts, splintered palm boles, empty condensed milk cans, shreds of nylon fishing net, the dull white sole of a training shoe.

I picked this last object out of the scum to look for a trademark which might give a clue to its origin. It was not a training shoe but the sole of a human foot, perhaps half an inch thick and trimmed of the toes. Its pulpy upper side, long since leached of blood and colour, was threaded with nematodes. On the underside were the callosities and scars of a life lived barefoot. Although it smelt I sat with it a short while, wondering when and how each scar had been acquired. It was quite broad and, despite sloughing, still deeply lined in the arch. Not a young person's foot. I imagined a middle-aged fisherman caught in the typhoon a few days earlier and presumed sharks had done the rest. It was hard to see why a shark would snip off the toes and leave the remainder or, indeed, how it could so cleanly have severed a flap of meat. But they are strange and beautiful creatures whose acute olfactory sense makes for impetuosity and abrupt switches of attention rather than thoroughness. Probably the victim had not been on his own and the animal had found a surfeit of food. I threw the sole back into the sea and rinsed my hands. Out under the waves would be sleek stomachs and powerful alimentary canals digesting a cigarette lighter, tatters of denim, a pair of spectacles.

III

MARGINALIA:
Seasickness

S ome people do not need actually to be in the water to experience the flexible boundary between life and death. For them it is enough to be on the sea in a boat. They are the victims of seasickness and frequently claim, while in the throes of this ailment, that they would far rather be dead. In this they are in august company. Cicero, having fled to sea to escape Mark Antony's sentence of beheading, was so seasick he gave up and returned to Gaeta, preferring execution to the unconsummated death sentence passed on him by the ocean. Given how long human beings have been seafarers, it must be one of the oldest forms of illness to be described, consistent in its symptoms from culture to culture as across the centuries. No doubt its cause was occasionally attributed to malevolent sea spirits or witchery, but even in Antiquity people were quite capable of being rational about it. Plutarch was curious as to why it occurred only on the sea and not on rivers. He blamed the smell of the sea and the apprehensiveness of the sufferer, perceiving the psychological component which seems to play a large part in the condition. Apparently he never made the connection between seasickness and motion.

There are several seventeenth- and eighteenth-century treatises on seasickness with titles like *Dissertatio de Morbo Navigantium*, for it must always have been recognised as a problem for navies as well as for ordinary travellers, and therefore worth serious medical attention. One suspects it only attracted more general and popular concern during the nineteenth century, when there were enough passengers taking part in the mass emigrations to the New World. By the time tourism proper started, especially with excursions from England to the continent in the latter half of the century, seasickness was the subject of dozens of booklets and articles. Most were more interested in remedies than in causes, the majority admitting that these were not well understood. Among the theories might be any of those suggested by doctors of the day:

(i)   an 'afflux of blood' to the spinal cord;
(ii)   disorientation caused by the rolling or heaving;
(iii)   'depression of the circulation';
(iv)   'displacement of the abdominal viscera';
(v)   the influence of 'changing impressions made upon the vision'; (obviously a fallacy, one writer remarked, since the blind are just as seasick);
(vi)   the influence of a 'marine miasma' or 'miasmatical intoxication';
(vii)   'sanguine congestion in the brain, provoked and entertained by the deranged centre of gravity';
(viii) 'centrifugal force within the blood vessels' produced by the oscillation of the ship.

As for treatment, this generally amounted to heavy sedation. Thomas Dutton, a popular medical author writing in 1891, presumably thought travellers also needed the placebo effect of a bizarre regimen. His 'cure' began a fortnight before sailing and consisted of a light diet, a digestive pill at night, a glass of salt water twice a week before breakfast and a four-mile walk daily. Three days before travelling his patients began taking a medicine of ammonium bromide and chloroform. Once aboard ship they were to reduce the bromide as far as possible, avoid ship's food and subsist on strong beef essence, dry biscuits and whisky or brandy and soda. In addition,

Dutton recommended any or all of the following: chloral hydrate (favoured ingredient of the Mickey Finn or 'knockout drops'), dilute prussic acid, iodine, amyl nitrite, cocaine in quarter-grain doses, creosote, cerium oxalate, soda bicarbonate, caffeine, eucalyptus and Nepenthe (a proprietary solution of opium in alcohol, dosage as per laudanum). The amyl nitrite was taken orally, diluted in alcohol. Any sufferer on this regime would be doing well if he was even aware of being on board a ship. Many travellers must have spent entire voyages in deep narcosis. In the meantime, starting a 'cure' two weeks before any possible onset of the ailment might be presumed to have the effect of almost guaranteeing seasickness. The sufferer had thoroughly prepared himself to be ill, whether from the sea or the prussic acid. The glasses of salt water are odd. They were no doubt emetic, but they might also have had a homoeopathic function, as if small doses of salt might make one immune to the briny. That might also go for the iodine, which was derived from seaweed.

Several Victorian experiments were made in which cabins, restaurants and entire passenger areas of a ship were mounted on gimbals so as to remain steady, but these were not a success. The engineering problems were considerable, the boundaries between the 'stable' and the 'moving' parts obviously being zones of great danger. Such things reflected the consensus that at the root of seasickness was the ship's motion. This was not quite as banally obvious as might seem. It had not occurred to Plutarch, after all, and until the mid-nineteenth century conditions on board most ships could induce sickness even if the vessel were tied up in port. The food was foul, the sanitation facilities fouler, and contagious disease easily transmitted in the cramped and generally overcrowded conditions. In the circumstances, any number of acute symptoms might mask or exacerbate those of ordinary seasickness. M. Nelken, in his advanced and sensible book *Sea-Sickness* (New York, 1856), added an appendix in which he gave details of the safety regulations made necessary by the flood of emigrants leaving Europe for America. The new regulations were an attempt to 'apply a remedy to the gross abuses which have caused such vast numbers of persons to be swept into the grave, during the few short weeks of transit across the ocean.' In 1848 Congress finally passed an Act which for the first time regulated the amount of space allotted to each passenger as well as their total number. Even before Nelken published his book these new laws had effected a dramatic change and 'already showed a great drop in the number of deaths aboard to an average of less than 1 per vessel'.

Nelken, like Dutton nearly forty years later, ascribed seasickness to motion, but unlike Dutton did not relate it to other forms of travel and perceived it as a special case of motion sickness. 'The same symptoms,' wrote Dutton, 'are often felt by some people, particularly children, when journeying by train or a vehicle, so we may have train-sickness, carriage-sickness etc.' This only needs updating with the addition of car sickness and airsickness. Airsickness used to be a much greater problem than it is now because earlier aircraft, like earlier ships, were smaller and lighter. They had only a limited ability to avoid or fly above bumpy weather. Nowadays, airsickness bags are sordid but touching relics of a bygone age. As such, they are similar to those few remaining drinking troughs for horses in towns and cities. Once in a blue moon one might be used, filling onlookers with curiosity and pleasure.

The latest medical thinking about seasickness, according to a naval surgeon in Plymouth, is agreed. It is caused by 'a discordant clash of information between the organs of balance and the eyes.' This diagnosis is charming, being wholly nineteenth century in its phraseology and vagueness. (It even begs the question of the blind.) Evidently the advances of the past one hundred years are all in the field of medication, though many people would vastly prefer cocaine to Kwells. One is left to wonder why, among other things, a clash of information in the head should provoke sickness in the stomach. Each of our two inner ears contains three semicircular canals which are set at right angles to each other in three planes. The canals are filled with liquid whose movement stimulates receptors in the ampullae at the ends of the canals. The swirling of minute particles of chalk suspended in this fluid generally enables the brain to maintain our balance in relation to a stable exterior world. When that world becomes unstable, medical theory suggests, the brain-as- computer (always a doubtful model for that organ) goes into overload, confused by too many variables and a surfeit of conflicting messages from the inner ears, the eyes, the soles of the feet, etc. Why some people should be more sensitive to this than others is unclear. It is not the same as wondering why some people are allergic to shellfish, because seasickness is not obviously a matter of pure physiology. Psychology clearly plays a significant role and so does habituation. Even poor sailors can in time acquire a degree of immunity or at least control if they have to. To most sufferers, however, the ailment remains like an advance symptom of death. It would surely be hard for such a person to view the sea at all impartially or shorn of its mortal associations.[1]

---

[1] 'Sensory conflict' theories of motion sickness are still the most favoured, although no neurophysiological evidence has yet been found to explain how or why the stimulation of brain regions dealing with spatial orientation should affect those that mediate nausea. A good deal of effort is now being expended on research into space motion sickness or SMS, which is merely the twenty-first century's version of the nineteenth century's preoccupation with seasickness. And the approach to alleviating the symptoms is scarcely different. Like Thomas Dutton before them, today's NASA doctors are reduced to trying a series of pharmaceutical blackjacks.

Ever since he failed to find the reef he thought he saw and the boat he was sure he had glimpsed, the lost swimmer has become conscious of the gulf he hangs over. At least the empty but navigable plain which surrounds him horizontally spreads itself beneath the sun's broad eye. Finding his way home again, back to life, will be a matter of simple luck or simple physics. A puff of wind here, an eddy there, and he will be reunited with his boat. If for a moment he were able to raise himself only fifty feet above the water he would spot it at once and the entire traumatic incident would be at an end.

Beneath him, though, lies a dimension which absolutely refuses to reduce itself to a matter of simple physics. The seabed is several hundred metres away—perhaps 1,000 if he is further from land than he thought. Up to a mile of water, in short. The swimmer tries to remember what a mile looks like. The entire length of Oxford Street, Centre Point to Marble Arch, but stood on end. As he contemplates this, something unseen like a gush of sepia roars soundlessly up at him from below, without warning, blotting out the sunlit layer which swathes him. This chill black torrent is overwhelming in its despair. It is as though a microscopic ghost had arisen from every test and skeleton of the uncounted radiolaria and plankton bedded on the bottom and had suddenly joined in a great upward fume. Far, far below, the basalt itself is calling in a language of aeons, and its empty message echoes up and spreads around him in a freezing, inky pool. This tectonic voice paralyses him. It mocks all human hope. It is worse than his first panic, worse even than the threat of death.

5

Deeps and the Dark

The wreck of the Florida in Jules Verne's *Twenty Thousand Leagues under the Sea* (1870), perfectly illustrating the enduring belief that any object will reach a level beyond which it cannot sink.

I

Why did Dr. Ballard, on his journeys to photograph the *Titanic*, habitually play classical music during the descent and rock music on the way back up? It is unthinkable that it could have been the other way around. The idea of 'the Deep' is so powerful that if we listen to the word as we say it a shiver may pass through in recognition of all the associations it has jarred into resonance. By comparison, 'heaven' is blank and thin, even faintly unserious. 'The Deep' is utterly solemn. Tennyson, whose childhood on the Lincolnshire coast left the recurrence of the sea and its imagery in his poems, knew the word's exact weight. Stately, funereal, mysterious, it spoke ultimately of loss: a steep dark bulk, time's liquid correlative which gulps down objects, lives, all that was and will be. Sometimes it is dreamily sinister, as when he layers the ocean horizontally to intensify sheer depth and discern his ageless monster:

> Below the thunders of the upper deep;
> Far, far beneath in the abysmal sea,
> His ancient, dreamless, uninvaded sleep
> The Kraken sleepeth . . .[1]

Elsewhere, he blurs the outline of a lost friend with the geological implacability of death:

[1] Alfred, Lord Tennyson, 'The Kraken' (1830).

There rolls the deep where grew the tree
  O earth, what changes hast thou seen!
There where the long street roars, hath been
The stillness of the central sea.[1]

Tennyson had read *Principles of Geology* and must have been struck by the passage where Lyell describes how 'many flourishing inland towns, and a still greater number of ports, now stand where the sea rolled its waves.'[2] These two themes— monsters and geology—recur over and over again in the intellectual life of the mid-nineteenth century as the question of 'the Deep' was finally tackled by science.

Alexander the Great allegedly had himself lowered into the Mediterranean in a glass cage whose door was fastened with rings and chains. He judiciously took food with him, anticipating a long vigil. It was a legendary business, suitable for enhancing the heroic myth, paralleled in the twentieth century by stories such as those told during the Chinese cultural revolution of Mao Tse-Tung strolling for an hour on the bottom of the Yangtse. In Alexander's case the things he saw were on a properly heroic scale. He observed a monster fish which took three days and three nights to swim past. Such was the insatiability of his scopic drive that, powerless inside his observation chamber, he nonetheless managed to include in his hero's gaze the subjects of another monarch, uninvited as he was to Neptune's abyssal kingdom. It was an exclusive as well as excluding view:

> None of the men who have been before me, and none of those who shall come after me upon the earth shall see the mountains and the seas, and the darkness, and the light which I have seen . . .'[3]

[1] Tennyson, *In Memoriam* (1850).
[2] Charles Lyell, *Principles of Geology*, 4th ed. (1835), Vol. I, p. 375.
[3] E. A. Wallis Budge, trans. of the Ethiopic version of pseudo-Callisthenes (1933).

The glass cage was all-important. Alexander was not in that humbler but more reliable alternative, a wooden barrel with a glass spyhole. Evidently he felt it important both to see and be seen, to be recognised as Alexander the Great by the creatures of the deep. Maybe this is a characteristic of heroes, for it is also told of him that after death his body was embalmed in honey and, according to his own instructions, exhibited in a glass coffin. This must have afforded an attraction of Leninesque proportions, and it would be interesting to know if he left orders that his eyes remain open beneath the honey the better to survey his awed pilgrims even as they him.

The legend of Alexander's descent makes plain that the depths of the sea is no place for ordinary mortals. It took a hero to confront it on anything like equal terms. Despite salvage activities, until the late eighteenth century the average European's mental image of the sea was literally superficial, of a navigable surface above an abyss. It was a treacherous surface, obviously, being liable to spasms of hostility or the unpredictable appearance of awesome creatures from below; but a seafarer needed to know only about winds, waves and currents, and the intentions of other seafarers. Anything deeper was hidden. Yet the work of early hydrographers, the attempts of Scandinavians to correlate the supply of fish with currents, and the demands of geologists to have their theories confirmed made it inevitable that the barrier of the deep sea would be tackled. It was a barrier in more than just the physical sense, however. There was something in the very concept of the abyss which paralysed thought. There seems no other way of explaining the long survival of certain fallacies more akin to superstitions, often promoted by scientists themselves in contradiction of their own laws. Two famous examples show this: the notion of the compressibility of seawater, and the temperature at which seawater reaches its greatest density.

By the end of the eighteenth century scientists knew perfectly well that water, unlike air, can scarcely be compressed at all. Even under great pressure the density of water changes little, certainly not enough to alter its viscosity by much. To the extent that it does change, temperature is a more important factor than pressure. Yet an extraordinary theory survived this knowledge, lasting well into the twentieth century. It held that as pressure increased with depth, seawater grew more and more solid until a point was reached beyond which a sinking object could sink no further. Thus, somewhere in the middle regions of the great abyss, there existed 'floors' on which objects gathered according to their weight. Cannon, anchors and barrels of nails would sink lower than wooden ships, which in turn would lie beneath drowned sailors who themselves lay at slightly different levels one from another, depending on their relative stoutness, the clothes they were wearing and, quite possibly, the weight of their sins. This notion was reflected in the old saying 'Jack will find his own level.' The popular belief was that having reached their level, bodies would forever drift and revolve in timeless suspension. After the *Titanic* disaster in 1912 it was reported that some of the relatives of drowned passengers had expressed dismay at the prospect of their loved ones wandering through the abyss in submarine limbo.

In the mid-nineteenth century this belief was held even by many scientists. Doubt had been voiced that the transatlantic telegraph cable would ever reach the seabed Lieutenant Maury claimed to have mapped. Might it not sink only so far, to lie conveniently above any ridges and crevasses? ('This would be to our benefit,' one nervous shareholder in the original Atlantic Telegraph Company wrote to a friend. 'Yet surely the conversants' [*sic*] voices, subjected to such uncommon compression, may emerge only as mouselike squeakings?,' thereby adding a further misconception of his own. Heady days for speculators, in both senses.) In an otherwise sensible book published in New York in 1844 we find:

> Heavy bodies, which will sink rapidly from the surface, do at length apparently cease to descend long before they have reached the bottom; the pressure of the water being such as to cause them to remain at certain depths, varying in proportion to their weights. Thus it is that the plumb line will not act beyond a certain length, and we have no means, of course, of extending our enquiries deeper.[1]

This passage embodies a strange and interesting idea which reverses conventional heuristic wisdom, namely, that theory can itself make experimentation impossible. At moments like this it becomes legitimate to think about a psychic barrier to exploring the deep. A similar implication, that there are certain things best left undone and certain places it is wiser to leave untrespassed, is no doubt behind the pseudo-scientific reasons periodically advanced for the impossibility of doing them and warnings of the disaster which must befall any attempt to breach the 'natural' limits to human activities. It had been predicted at much the same period that speeds above that of a galloping horse would necessarily kill railway passengers. (In the twentieth century, too, ideas were dreamed up to show how any attempt to travel in space would be doomed. Arthur C. Clarke once quoted a man who had written in the 1950s to inform him there was a barrier separating the outer atmosphere from space proper, an 'adamantine membrane' which kept our air in. Anyone having the temerity to force his impudent way through this protection put there for our benefit by A Being Wiser than Ourselves would risk being sucked at infinite speed into outer darkness, followed by all the planet's air.)

---

[1] Anon., *The Ocean, A Description of the Wonders and Important Products of the Sea*, p. 17.

The anxiety behind such misgivings has accompanied all technological advance. Why else the notion of the sound 'barrier'?
Those who worry about 'breaking' the speed of sound or
infringing the depths of space are hardly distinguishable from
those who, a century earlier, believed the depths of the sea
could never be explored.

Any object or creature floating on the sea's surface is
already supporting with its body the weight of a column of
atmosphere tens of miles high, a pressure defined as *one atmosphere*. To a creature accustomed to sea-level pressures, such as
most human beings, this is not noticeable since his body will
be in equilibrium with atmospheric pressure. The moment he
descends below the waves, however, he will carry the additional weight of a column of water which, being so much denser
than air, bears very heavily on his lightly pressurised body.
Water pressure increases by an entire atmosphere for every ten
metres of depth. Although the human body is seven-tenths
fluid, its pockets of air, its cavities and the materials and construction of many of its components make it far less dense than
water. In order to prevent himself being squeezed to death
beyond a certain, quite shallow, depth, a human needs to be
protected by an outer casing (a diving suit or a submersible)
which can resist this external pressure and permit an enclosed
environment at his preferred sea-level pressure of one atmosphere. Clearly, the deeper he goes, the stronger this protective
cell will need to be. The pressure at the bottom of the Marianas
Trench is some 1,170 atmospheres, where each square inch of
the seabed, or of a body lying on it, bears a weight of 7.75
tonnes.

The assumption that water itself could be squeezed 'solider'
by such pressures, as a human body would, no doubt derived
from *Homo*'s habit of seeing the physical world in his own
image. This fallacious idea that water could be compressed into
an impenetrable layer somewhere 'below the thunders of the

upper deep' was remarkably tenacious although it never stopped serious attempts to take soundings at ever greater depths. Certain scientists did wonder whether the sounding line might not really be going any lower but simply be piling up in loose coils on this invisible floor like a thread of honey falling on to a plate. Samples of mud brought up were explained as the sediment which had likewise collected there over the course of millennia, presumably building up into a false bottom to the sea. It was the second misconception, however—about the temperature at which seawater reaches its maximum density—that was the more seriously and widely held and had further-reaching effects on early oceanography. It was more baffling, too, since it was blandly assumed—and could very simply have been disproved—that salt water behaves like fresh and is at its densest at 4°C. Strangest of all, the actual temperature had already been established, as Wyville Thomson later pointed out. 'In 1833 it was ascertained that the temperature of sea water at its maximum density is—3.67°C, and even before that it was known that sea water can be colder than the freezing point of fresh water and still remain liquid.'[1]

Unfortunately, the reluctance to accept what had already been scientifically determined was compounded by the inadequacy of the instruments of the day, as Sir James Clark Ross unwittingly showed on his Antarctic expedition of 1839–43. From the decks of HMSS *Terror* and *Erebus* (which only two years later were to disappear famously when Sir John Franklin took them to the Arctic to search for the north-west passage), thermometers were lowered on sounding lines deep into the South Polar ocean. When pulled up, those that had not imploded under the pressure read 4°C. What Ross did not know was that in those latitudes the water temperature drops about 1°C every 550 fathoms, a decrease which by sheer mis-

[1] C. Wyville Thomson, *The Depths of the Sea* (London, 1874).

chance happens exactly to compensate for the opposite effect of pressure on an unprotected thermometer. In point of fact his own uncle, Sir John Ross, had already found a temperature of –1.8°C more than twenty years earlier with a thermometer which must have been protected against pressure. This was in 1818, during the expedition to Baffin Bay in HMS *Isabella*. Sir John not only established this temperature at a depth of 1,050 fathoms but he also brought up a beautiful 'Medusa's Head' starfish (a basket star) from the same depth, something which was conveniently overlooked in the following decades.

The prevailing view of the abyss was now of a vast body of water at a uniform temperature of 4°C, unmoved by either winds or currents. (Had the temperature been allowed by theory to vary between 4°C and –3.67°C, slow convection currents would have been set up.) Without movement, scientists reasoned, there could be no circulation of dissolved oxygen and no renewal of any food particles in suspension. This in turn would ensure the abyss was 'azoic,' or lifeless, since a stagnant body of water under huge pressure, at barely above freezing point and utterly without light, could not conceivably support life.

The word 'azoic' was coined by Edward Forbes, who in the 1840s tried to discover where the boundary lay between the upper part of the ocean which would support life and this great 'lifeless' zone. Forbes was a Manx naturalist who in 1842 sailed to the Aegean in HMS *Beacon* to study the vertical distribution of marine animals. What he found confirmed his theory to his own satisfaction and he came home saying he considered 300 fathoms to be the absolute limit of animal life in the ocean. In fact, as Margaret Deacon has pointed out, such life is particularly sparse at that depth in the Mediterranean.[1] Forbes also went dredging in the Firth of Forth, taking with him the young Wyville Thomson who for a long time accepted

[1] Margaret Deacon, *Scientists and the Sea* 1650–1900 (1971).

his mentor's 'azoic' theory and thirty years later would write shamefacedly:

> We had adopted the current strange misconception with regard to ocean temperature; and it is perhaps scarcely a valid excuse that the fallacy of a universal and constant temperature of 4°C below a certain depth . . . was at the time accepted and taught by nearly all the leading authorities in Physical Geography.[1]

If it caused a scientist shame in the 1870s to recall the cant of his youth, it is not easy to know how to treat an episode which took place in the House of Commons nearly a century later. On 12 April 1961 the MP Hector Hughes asked the Civil Lord of the Admiralty, Ian Orr-Ewing, certain questions about recent 'experimental missions' beneath the Arctic ice cap by the Royal Navy submarines *Finwhale* and *Amphion*. The Civil Lord would give no details, explaining 'It would not be in the national interest.' His questioner shifted to the less classified ground of schoolboy physics and the following exchange took place:

> *Mr.. Hughes:* Can the hon. Gentleman say why the water under the North Pole does not freeze while the water on the surface of the North Pole does freeze? . . . Is the water under the ice kept warm by the heat generated from the centre of the earth?
>
> *Mr. Orr-Ewing:* In view of the hon. and learned Gentleman's interest in bathing, I can understand his anxiety about where the ice forms. If he studies the physical tables, he will find that the water is most dense at 4 degrees centigrade and rises to the surface when it reaches 0 degrees centigrade and starts to freeze.[2]

[1] Wyville Thomson, *The Depths of the Sea* (1874).
[2] *Hansard*, Vol. 638, p. 235.

The 'azoic' theory held into the second half of the nineteenth century. When HMS *Bulldog* resurveyed the transatlantic route for a telegraph cable, soundings were taken down to 2,000 fathoms. When the sounding lines brought up starfish, everyone maintained that the creatures must somehow have become entangled as the lines were being pulled through shallower levels. It was only in 1860 that the theory was finally and unequivocally exploded when a section of telegraph cable was fetched up for repair off the coast of Sardinia. The cable had been laid three years earlier in more than 1,000 fathoms of water (i.e. over a mile deep) and it was found that various marine animals were encrusted on it, their anchoring filaments having worked their way into the outermost layer of insulation. It was quite impossible to argue that they had 'become entangled'; they had quite evidently grown there. Thus it turned out that the abandoning of the 'azoic' theory happened neatly to coincide with the publication of Darwin's theory of evolution.

Practically overnight the almost universal conviction that the deeps were sterile changed to intense speculation that they might actually conceal life forms as well as mineral wealth. As Wyville Thomson was to observe, 'the land of promise for the naturalist . . . was the bottom of the deep sea.' It is perhaps hard now to imagine the ferment which the scientific method was causing in the middle of the nineteenth century. In the 1860s wild speculation became common when the abyss was considered in the new Darwinian light which made necessary a complete revision of prevailing ideas about the planet's history, about man's position in the 'natural' world and about his relationship to a 'creator.' Two fields of study, geology in general and the fossil record in particular, had an especial bearing on oceanography. In his *Principles of Geology* Charles Lyell had avoided tackling head-on the six-day, Genesis version of creation. Instead, he confined himself to pointing out that the Earth's features could all quite adequately be explained in terms of the simple physical processes which were visibly still shaping it: tension/compression and erosion/sedimentation. This was elegant and satisfactory. The implications might have raised some eyebrows but few hackles since it concerned only the inanimate world of petrology. The real furore was to come a quarter-century later when Darwin made man and the animals also subject to an evolutionary process which led to notions of trial and error, sports and dead ends, casual extinctions, uncomfortable family connections and—worst of all—to the logical conclusion that *Homo*, far from having been perfected as Nature's last word, must himself still be evolving. Darwin's theory also made plain the crucial evolutionary role played by environment. Where conditions were (in geological timescales) fickle and changing rapidly, the species that survived were those which best adapted and evolved to keep pace with them. It was this idea which, coinciding with the demise of the 'azoic' theory of the deeps, generated speculation as to what kind of creature might have adapted itself to conditions hitherto considered inimical to life. All the factors which until so recently had indicated sterility—absence of light, intense cold and pressure, no movement—now suggested the one place on Earth in which to look for unmodified ancient creatures, 'living fossils.' (The terrestrial equivalent was the search for the 'missing link,' a hypothetical extinct creature midway between the anthropoid apes and man. Storybook quests such as Conan Doyle's *The Lost World* grew directly out of the notion that living fossils might yet be found on dry land.)

It was as if Darwin's intellectual leap had caused natural laws to rewrite themselves and the invisible 'floors,' which until so recently had prevented sounding lines from reaching the deep ocean bed, all collapsed at once—like adamantine membranes—and began letting through an array of plummets, grabs, dredges, corers and other sampling devices. Now dredging expeditions began finding sea animals which resembled fossils. Specimens of a stalked crinoid, *Rhizocrinus lofotensis*, were brought up from the deep off Norway. No known modern coastal species of this sea lily had a stalk. This was followed by all kinds of hitherto unknown varieties of starfish and sponges from the world's oceans and further reinforced the idea of 'living fossils' which would presumably be more and more archaic the further down they lived. Though completely wrong, this notion did at least give oceanography the last impetus it needed to begin the systematic exploration of the deep.

Now that the problem was no longer conceptual, the major difficulty lay in designing equipment for taking deep soundings and samples. Where sounding was concerned, it was one thing to pay out a weight on the end of a line over the side of a ship but quite another to know when it had reached the bottom, since even the thinnest sounding wire weighed a lot with 2,000 fathoms deployed. Men became expert at judging when the plummet had stopped, keeping a sensitive finger on the line as it vanished overboard. However it was done, it was a laborious process. Thomson recorded that in 1868 aboard the *Lightning* a 'Hydra' sounder was used which, weighted with 336 pounds, took 33.5 minutes to reach 2,435 fathoms off Biscay and two hours and two minutes to heave back up again with a few ounces of grey Atlantic ooze. This system was much modified in detail but little changed until the invention of sonar depth-sounding. Even forty years after Thomson, the young Boyle Somerville aboard the *Penguin* was using a more or less identical process. 'Birmingham Wire gauge no. 20, galvanised' was paid out over a nine-inch diameter wheel from a huge drum holding about 6,000 fathoms. The little wheel was connected to a counter graduated in fathoms. There was a complex system of inertia brakes acting on the big spool, automatically gripping and releasing it according to the ship's motions, thereby maintaining a safe and even tension in the wire. The weight on the end was called a 'driver rod,' actually an iron tube, and was supplemented by two cone-shaped lumps of cast iron. Despite the brakes on the drum the entire process had to be watched 'like a hawk,' and in fact they lost 10,000 fathoms (nearly 11.5 miles) of wire, two driver rods and two deep sea thermometers before they were successful. Then, 'We were the first to see land that came from a depth below sea-level which was just a little more than the height of snow-topped Everest is above it.'[1] The *Penguin*'s skipper had been on the *Challenger* with Thomson and related a story which showed that where certain things were concerned, oceanographers had from the beginning had a sense of the priorities. At first, he remembered, the scientists had attached bottles of beer to their sounder in order to cool them. When they came up again from the icy depths the seals were intact and the corks still in place but the contents found to be 'the very best sea-water.' This method having failed, the beer was set to cool in the samples of ooze dredged up from 2,000 fathoms. This was very cold, about 35°F, and no scientific investigations were made of the sample until the beer had reached its optimum coolness.

Pranks aside, that earlier expedition had marked the high point of nineteenth-century oceanography. At Christmas 1872 HMS *Challenger* had sailed from Portsmouth for what turned out to be a three and a half year voyage. At the time it was the best-equipped (at government expense) scientific expedition ever mounted. Some would argue it remains the greatest of all such voyages of discovery. One of the expedition's specific hopes was to find 'living fossils,' and the scientists aboard vainly sifted ton after ton of bottom samples in search of wriggling trilobites. What they did find was life in even the deepest parts of the ocean. The 'azoic' theory was by now officially dead, of course; yet it lingered on in vestigial form owing to the technical inadequacy of the sampling instruments of the day. That is,

[1] Boyle Somerville, *The Chart-Makers* (1928).

the expedition's director, Charles Wyville Thomson, observed there was life both at the top and the bottom of the ocean for the simple reason that it was sustainable there. Even the deepest sediments were colonised by organisms such as worms, echinoderms and omnivorous crustaceans, so it was not surprising that abyssal and even hadal (the deepest of all; that is, over six kilometres) ecologies could also support highly specialised types of fish. Yet Wyville Thomson still felt sure that the oceans' middle layer would turn out to be sterile because it lacked nutrients. Such particles as there were fell straight through it and down to the seabed. His problem lay in proving or disproving this. There was as yet no reliable way of taking samples from these intermediate zones without also catching specimens from the upper layer through which a net had to pass twice.

During this long voyage there came a reminder of another enduring myth, and in sad circumstances. William Stokes, a young sailor aboard *Challenger*, was killed in an accident on deck. On the day of his burial at sea, a delegation of his shipmates approached Wyville Thomson and enquired anxiously whether their friend's body, when suitably weighted, would truly reach the bottom or, as tradition had long maintained, would float at some indeterminate depth. Wyville Thomson was able to reassure them that his remains would indeed reach the bottom. A sounding taken shortly before Stokes's funeral read nearly 4 miles, at that time the deepest ever measured.

What would happen to the boy's body on its long fall of over 21,000 feet? A two-pound cannon ball would take well over half an hour to reach the bottom. A corpse, far less dense and streamlined, might take hours, assuming it was not attacked and dismembered on the way down. Just as it is impossible at any funeral entirely to suppress anxiety and not wonder, in however fleeting and censored a fashion, exactly what the worms or flames will shortly do, there must have been scientists on deck that day wondering about the effects of pressure on the late William Stokes. (It is most likely that a human body has never been retrieved from such a depth. Although corpses must have been subjected experimentally to enormous pressures to see what happens, the results are presumably buried in the files of naval research institutions.) Wyville Thomson had already written in musing manner: 'At 2,000 fathoms a man would bear upon his body a weight equal to twenty locomotive engines, each with a long goods train loaded with pig iron.'[1] By now he had got his facts straight about the incompressibility of water.

> Any free air suspended in the water, or contained in any compressible tissue of an animal at 2,000 fathoms, would be reduced to a mere fraction of its bulk, but an organism supported through all its tissues on all sides, within and without, by incompressible fluids at the same pressure, would not necessarily be incommoded by it. We sometimes find when we get up in the morning, by a rise of an inch in the barometer, that nearly half a ton has been quietly piled upon us during the night, but we experience no inconvenience, rather a feeling of exhilaration and buoyancy, since it requires a little less exertion to move our bodies in the denser medium.[2]

At some point the air-containing parts of Stokes's body would have ruptured, principally those of his face, chest and abdomen. The head would not have burst because the cranium contains no air, only incompressible liquids, but the delicate bone honeycombs of his sinuses probably collapsed before water could leak in to equalise the pressure. Sooner or later the chest could have imploded, the broken ends of the ribs coming through the skin. Any air in the gut would probably rupture the abdomen, so if Stokes had been a flatulent boy

---

[1] Wyville Thomson, *The Depths of the Sea* (1874).
[2] *Ibid.*

it would in the end have been his literal undoing. The pressure would also have been likely to cause stress fracturing of certain parts of his skeleton. There might, for example, have been some splitting around the pelvic crest since the abdominal wall is highly compressible whereas the pelvis is not. The same would have applied generally to any structures of finely divided bone (i.e. not solid and thick as in the femur). Stokes would have arrived on the bottom somewhat smaller than he had been on the surface, especially if he was fat, since fat is more compressible than water. The creatures of the seabed would make short work of his flesh, of course, once they had found their way through the holes his rib-ends had poked through the canvas; yet even his skeleton would not last as long as in a conventional earth burial since bone softens in seawater as its salts are leached out by osmosis. Thus softened, the boy's remains would have crumbled away beneath the pressure.

A vivid demonstration of what deep-sea pressure can do is shown in the experiment beloved by modern oceanographers of sending down with a piece of high-tech equipment an ordinary empty polystyrene coffee cup. It comes back in miniature, a tiny white thimble, all its insulating air cells having collapsed. Yet there seems to be a reluctance to perform this experiment with the body of an animal. I scoured the *Farnella* for a ship's rat, hoping that if we could kill a brace we might send them down a couple of thousand fathoms to see what ruptured, but this piece of curiosity was greeted with cries of distaste and accusations of being a ghoul.

In all, the *Challenger* covered 68,930 nautical miles and at the end of three and a half years brought back so many samples of marine plants, animals, seawater, sediment dredgings and corings that it took the next nineteen years to process them. By then Wyville Thomson was dead and his place had been taken by his assistant John Murray. The subsequent report, which by 1895 had reached fifty volumes, has been described as 'the most complete expression of man's knowledge of the deep sea'.[1] Perhaps as importantly, the enterprise encouraged similar expeditions by other nations, principally the USA, France, Germany, Russia, Italy and the Scandinavian countries. Even Monaco came to hold an honourable position in marine research since Prince Albert I was himself an expert yachtsman and oceanographer who financed his own expeditions. Among his most valuable contributions was a collection of specimens from the intermediate zone which Thomson had thought might be azoic.

Part of the *Challenger*'s achievement was to have laid to rest various misconceptions and to have settled theoretical disputes. Prominent among the latter were post-Darwinian issues concerning living fossils and the Earth's geological evolution. The short answer to the expectation that the deeps concealed living fossils was that they did not. What they revealed was absolute proof that even the greatest depths were neither immobile nor sterile, and that they supported species which, far from remaining unchanged for sixty million years or more, had evolved their own range of special adaptations.[2] In the

---

[1] Susan Schlee, *A History of Oceanography* (1975). I am much indebted to this excellent work for many details in this and other chapters.

[2] The first time a deep-sea creature was found to fit the fervid category of 'living fossil' was in 1938 with the catching of the first coelacanth. There are good reasons for disliking the whole notion of 'living fossils,' some of which have been noted by Stephen Jay Gould with reference to horseshoe crabs. In taking issue with the meliorist view of evolution and with the tyranny of conventional iconographies –trees and ladders—he objects most to the idea that 'old' necessarily means 'primitive' or 'simple,' as if always to imply the superiority of *Homo sapiens sapiens*. In addition he says, 'We mistakenly regard horseshoe crabs as "living fossils" because the group has never produced many species, and therefore never developed much evolutionary potential for diversification; consequently, modern species are morphologically similar to early forms' (Stephen Jay Gould, *Wonderful Life*, 1990). Where the modern coelacanth is concerned it cannot be considered a 'living fossil' because no other members of the species *Latimeria chalumnae* have ever been found as fossils. Come to that, 'no other species assignable to the genus *Latimeria* has been found as a fossil either' (K. S. Thomson, *Living Fossil*, 1991).

mean time, other professional judgements were painfully exposed as incorrect. A few years before the *Challenger* expedition set sail Darwin's friend and champion, T. H. Huxley, had formulated two hypotheses which became *causes célèbres*, one concerning a notional living fossil of the most primitive kind, the other geology.

In June and July 1857, HM frigate *Cyclops*, while sounding down to 2,400 fathoms, brought up some sediment samples which in due course arrived back in London for examination. Huxley, at the time Palaeontologist at the London School of Mines, had them preserved in strong alcohol and appeared to forget about them until 1868. On re-examining them after eleven years he found a transparent jelly and became convinced that this was a living slime which carpeted the deep ocean floor, ingesting ooze and forming a rich layer of protoplasm which became a food supply for other life forms. As such, he gave it the name *Bathybius haeckelii* in honour of the German biologist Ernst Haeckel. Haeckel had been much struck by Darwin's theories and was preoccupied with finding a primitive organism which might provide the missing link between inanimate matter and life. The catchphrase of the day was 'abiogenesis' or 'spontaneous generation,' to describe the belief that living organisms could develop from non-living matter. (At a microbiological level this was exploded by Pasteur. In the present century an updated form of the idea was floated when amino acids, the 'building blocks of life,' were generated in the laboratory by imitation lightning discharges in mixtures of gases thought to approximate Earth's primordial atmosphere.) Haeckel was convinced *Bathybius* was the basis of all evolution, the original living matter or *Urschleim*. Whatever else, it would have to be a subject for investigation aboard *Challenger* since once Huxley had found and named it everybody else seemed to be dredging up samples and it was important to establish whether this primordial slime could be found evenly distributed throughout the world's oceans.

For two years *Challenger* found no *Bathybius* and finally the expedition's chemist, John Buchanan, discovered that he could reproduce this jelly-like substance when he preserved bottom samples in alcohol. He came to the conclusion that this famous protoplasm was no more than calcium sulphate precipitated out of seawater by the alcohol. Thomson at once wrote to Huxley who promptly, and with immense dignity, admitted the correctness of this chemical explanation and his own error. From that moment *Bathybius* was dead, although several scientists tried in vain to discredit the explanation and its discoverer's retraction.[1]

The second of Huxley's hypotheses also concerned deep-sea ooze, but in a geological rather than biological context. This was his theory of the 'continuity of chalk' which, briefly, stated that deep-sea ooze turned into the chalk deposits found on land, so that the continents were formed of the compacted material of the seabed. The essence of this position was the belief of the day that land could only move vertically up and down, for this was long before men like Emile Argand and Alfred Wegener had proposed lateral movement and continental drift. This gave rise to a debate between those scientists who believed the ocean basins and the high continents slowly traded places, and those who thought basins remained basins and continents continents. The *Challenger* soon put paid to the

[1] An interesting postscript has recently besen added to the *Bathybius* story by Dr. A. L. Rice at IOS, suggesting that the seasonal nature of its original collection implies that some of it, at least, could have been detritus of the spring phytoplankton bloom forming a light, flocculent 'fluff' on the seabed. This would explain why the *Challenger* failed to find samples since, being the marine equivalent of thistledown, it simply puffs out of the way of dredges and epibenthic sledges. (See A. L. Rice, 'Thomas Henry Huxley and the Strange Case of *Bathybius Haeckelii* . . .,' *Archives of Natural History* 2, no. 2 (1983, pp. 169–80).

continuity of chalk theory, too. It found that deep-sea oozes were quite distinct chemically from rock formations on land. Besides, geologists had for years been turning up shallow-water fossils in chalk beds on land, showing they could never have been formed in the deep oceans. The upshot was that no evidence was found either for drowned continents or rising ocean basins. This was a particular disappointment to palaeontologists, zoologists, botanists and others who thought they needed a sunken land bridge to explain how they were finding close correspondences between species of animals and types of geological formation in otherwise widely separated land masses. They, too, had to wait for Wegener as well as for theories of convergent evolution which explained how unrelated organisms can evolve similar shapes and adaptations in response to similar environments.

Behind this gathering of knowledge and the dispelling of misconception and superstition grew a desire to visit the deeps in person. In a way, the development of the submarine from the early twentieth century onwards was merely tantalising, since submarines were incapable of descending deeper than a few hundred feet, scarcely beyond the euphotic zone. Compared with reaching the deeps, this was the equivalent of getting a toe wet. Besides, submarines were war machines, not research vessels, and were far too big. They had the inherent problem of needing to support a large volume of air against great external pressure. This could only be achieved by massive construction, otherwise they would be crushed like a ribcage. Not until the 1930s did a true hero emerge prepared to put his trust in a piece of equipment which Alexander the Great would certainly have disdained.

This was the bathysphere, a name coined by its American inventor, William Beebe. The concept was simplicity itself. A thick-walled steel sphere with a circular entry hatch and a tiny porthole would be lowered like a plummet over the side of a ship on the end of a cable. In effect, it was an eyeball on a string. It could do nothing of itself but carry man's sight into unseen regions. Beebe's accounts of those early dives are in a sense laconic, even though shot through with terrifying images of the physical forces involved. The bathysphere was once sent down empty on a test dive and when hauled up was much heavier than usual. As the first bolts of the hatch were loosened needle jets of water sprayed out, showing it was partly full and under great pressure. It was clear to everyone that at some point in the loosening process the entire hatch might blow off, yet Beebe and his companion Otis Barton went on unscrewing the nuts with spanners while standing as far to one side as they could. When it finally did blow the heavy steel plate missed them both by fractions of an inch, flew the length of the deck, humming, and dented a donkey winch. Both men eventually felt the technological problems had been mastered, however, and made a historic series of dives cramped for hours in the tiny space, taking it in turns to squint awkwardly out of the peephole and dictate via telephone what they saw, while a female colleague in the ship far above took it all down in short-hand. A photograph taken on deck of Beebe emerging from the bathysphere shows the physical toll these long dives took. He is barely able to get through the tiny hole, so stiff is he with cold and cramp. There is no clutter of emergency equipment on deck, no officious bustle of rescue teams dressed in special gear; just a thin man with a lined face wearing slacks and canvas shoes being helped out of a steel ball which looks not much bigger than a large mine. In 1934 he and Barton reached the record depth of 3,028 feet off Bermuda.

When Beebe's accounts were not laconic they were filled with the excitement of a man who knows he is seeing things which no other human has ever seen and who responds with the keenest aesthetic pleasure. 'If one dives and returns to the surface inarticulate with amazement and with a deep realization of the marvel of what he has seen and where he has been,' Beebe wrote, 'then he deserves to go again and again. If he is unmoved or disappointed, then there remains for him on earth only a longer or shorter period of waiting for death . . .'.[1] Beebe and Barton did go again and again. After their record descent, Beebe observed, 'When once it has been seen, it will remain for ever the most vivid memory in life, solely because of its cosmic chill and isolation, the eternal and absolute darkness and the indescribable beauty of its inhabitants.'[2] He was particularly attentive to colour and the changes associated with depth. As the bathysphere was lowered through the fathoms Beebe relayed the vanishing of the comforting, warm rays of the spectrum as the colours from red through yellow to green were progressively filtered out, leaving the rest to 'chill and night and death.' He tried to describe what was left, a paradoxical and strange illumination which was both twilight and brilliant.

> It was of an indefinable translucent blue quite unlike anything I have ever seen in the upper world, and it excited our optic nerves in a most confusing manner . . . the blueness of the blue, both outside and inside our sphere, seemed to pass materially through the eye into our very beings.[3]

I quote Beebe because he is both scrupulous and imaginative, and his text is full of small observations which anyone who is thoughtful about the sea will immediately recognise as authentic, such as that you don't get wet when you dive, only when you surface. He was much taken by the luminous fish that swam past his window and lamented how much more he must be missing. Alexander the Great, scopic prodigy that he was, had watched a fish so huge it had taken three days to pass.

[1] William Beebe, *Half Mile Down* (New York, 1934).
[2] *Ibid.*, p. 175.
[3] *Ibid.*, p. 109.

Beebe merely records sadly, 'A gigantic fish could tear past the window, and if unillumined might never be seen.'[1]

Years later in 1949 Otis Barton made a descent off California which increased the world depth record to 4,500 feet, but the day of the bathysphere was done. The next step was taken by Auguste Piccard in his bathyscaphe. This device, which gave him the independence of not having to dangle helplessly on a hawser from a mother ship, was the undersea version of the balloons in which he had been setting world altitude records at the time Beebe was making his first pioneering descents. The bathyscaphe consisted of a pressurised chamber not unlike the bathysphere but slung beneath a large, lightly built tank full of petrol. Since petrol is lighter than water this was the equivalent of a gas envelope, and because the contents were incompressible there was no need for great strength and weight. Ballast took the bathyscaphe down and, once that had been released at the required depth, the flotation chamber brought it back up again. Piccard emphasised the ballooning pedigree by naming his first bathyscaphe the *FNRS 2*. (The original *FNRS* was a balloon named for the Belgian National Fund for Scientific Research which had supported the project.) Piccard's account is more technical than Beebe's, mainly because the engineering problems he had to solve were far more complex.[2] Although its manoeuvrability was very limited, the bathyscaphe was untethered and independent of a mother ship. It might very easily have stuck on the bottom without the remotest hope of rescue, particularly if its flotation tank were holed. Its inventor's story is a triumphant record of doggedly surmounting each new technical problem as it arose. His ingenious answer to the question of how to jettison ballast was a case in point. He needed to be able to dump weight in accurately controlled amounts but the pressure outside the capsule precluded any mechanical device pass-

[1] *Ibid.*, p. 221.
[2] Auguste Piccard, *Au fond des mers en bathyscaphe* (Paris, 1954).

ing through it. Apart from the additional danger of leaks, boring holes in the steel shell (which was cast and milled in two hemispheres) would weaken it. His solution was to use steel instead of lead shot as ballast and to jettison it through a separate chute around whose circumference were electromagnets. By pressing a switch Piccard could energise the magnets outside and lock the balls solid, blocking further release.

The bravery of these men seems extraordinary now, and it would be churlish to complain that Piccard was no Beebe when it came to describing what he saw when he went down. The fact is, his brilliantly engineered invention took him down very much further. In 1960 the *Trieste*, the latest version of the original bathyscaphe, reached the bottom of the Marianas Trench at 10,916 metres, some 35,800 feet or better than 6.75 miles. This is as deep below the ocean's surface as the highest-flying passenger aircraft leaving its white contrails is above it. Effectively, it was the deepest point in the oceans. Quite possibly there are places where this is exceeded by a few metres,[1] but to all intents and purposes man had gone as deep into the oceans as he ever would, just a century after Darwin's *The Origin of Species* was first published. Since then, the technology of deep-sea descent has become ever more refined and flexible, permitting proper, if limited, exploration. As with air travel, systems have become very much safer. This is by no means to belittle the courage of men like Robert Ballard, since the possibilities for disaster are still endless, miles beneath the last glimmers of daylight and with prodigious pressures ready to slam shut the tiny bubble of living space at the first sign of a weakening rivet.

The pleasure Beebe took in the luminous fish he saw was mixed with wonder at this evidence of a rich and entirely alien

---

[1] In 1984 the Japanese survey vessel *Takuyo* used a multibeam echo sounder to record an extreme depth of 10,924 metres in the Marianas Trench.

way of life where ambient darkness was as little a problem as either cold or pressure. More than sixty years earlier when aboard *Porcupine*, shortly before the *Challenger* expedition, Wyville Thomson had noted the light given off by coelenterate fauna such as gorgonians and sea pens brought to the surface in the trawl, marvelling that it should be bright enough for him to be able to read his watch by it. On one occasion what he saw gave him a glimpse into that illuminated underworld.

> The trawl seemed to have gone over a regular field of a delicate, simple Gorgonid . . . The stems, which were from 18" to 2 ft in length, were coiled in great hanks around the beam-trawl and engaged in masses in the net; and as they showed a most vivid phosphorescence of a pale lilac colour, their immense number suggested a wonderful state of things beneath—animated corn-fields waving gently in a slow tidal current and glowing with a soft diffused light, scintillating and sparkling on the slightest touch, and now and again breaking into long avenues of vivid light indicating the paths of fishes or other wandering denizens of their enchanted region.[1]

Much research has gone into the bioluminescence of different marine organisms, a subject made more complicated because the light has no unitary function. It seems that flashes of luminescence may be used variously as a defence, to entice prey, and as a sexual display. It may be seasonal or constant. It is even thought that some creatures may adroitly vary the wavelengths they emit, thereby using light itself as a method of camouflage. This is the same principle as the red colorations used by deeper reef creatures to make themselves look grey and stonelike, though beyond a limited depth there is little point in talking in terms of colour. The theory goes that an animal could camou-

[1] C. Wyville Thomson, *The Depths of the Sea* (1874).

flage itself by emitting low levels of light if it exactly replaced that lost by absorption on its upper surface. The light would have to be of precisely the right strength, at the right wavelengths and of the right angular distribution (since below about four hundred metres the remaining light falls vertically and is no longer refracted at other angles). There is increasing evidence to support this theory, and certainly the eyes of many creatures of the deep twilight and lower zones are highly sensitive to light and to the subtlest variations in its intensity and wavelength. Animals such as squid and hatchet fish use amounts of daylight which would appear indistinguishably black to human eyes in order to regulate their vertical migrations.

The discovery that vast numbers of animals rise to the upper waters at night and return to the depths during the day was surrounded by secrecy in World War II. Three scientists experimenting with sonar aboard the USS *Jasper* in 1942 had found a layer in the water at between 1,000 and 1,500 feet from which echoes bounced as if it were solid. This was not made public until 1946 because it was thought an enemy submarine might take advantage of the layer by hiding beneath it. In 1945 the Scripps Institution of Oceanography found that this layer moved up at night and down during the day and concluded it must be alive. Now known as the Deep Scattering Layer, its movement varies seasonally and from place to place. It consists of huge numbers of small animals migrating punctually up and down the water column, some by as much as 1,500 or 2,000 feet. The DSL is probably the chief cause of bogus sonar contacts, and many a ship has reported 'lost' land lurking just beneath the waves where later investigators have found only thousands of feet of water.

The deeps triumphantly disclosed the consequences of Darwin's ideas of natural selection in that the often bizarre colours and shapes of abyssal fauna emerged as exquisite adaptations to extreme circumstances. The 'azoic' theory had betrayed as nothing else the limits of understanding of the nature of life, and how erroneous all judgements were when based solely on human considerations of what might constitute a liveable environment.

Where 'monsters' are concerned, they may yet be found, although it is unlikely, owing to scarcity of food, that they will be from the very deepest parts of the ocean. However, if the Kraken is mythical, the giant squid is not. Huge specimens— and fragments of even huger—have occasionally surfaced. It is clear from measuring the sucker marks on dead whales that immense battles must take place in the middle deeps involving squid of a size never yet seen. The great mass of the oceans remains unexplored, even as the contours of their beds are electronically surveyed. Their waters must hide many species strange to taxonomy, but this is hardly surprising. Letting down nets here and there may catch few creatures with acute sensory equipment and evasive powers. The world a mile or more down keeps its secrets well, with neither victors nor victims necessarily leaving the least trace of their lives. As regards the deepest trench faunas, there has been relatively little recent research because most of the effort and money has been directed towards studying the vent communities around 'black smokers,' which have the required glamour to attract funding.

As to geology, the seabed turns out to be of great use in climate modelling. It is possible to weigh the atoms of oxygen trapped in fossil shells brought up in sediment cores and determine what the temperature was when the creatures were alive. Such cores have also yielded information about monsoons and glaciation. It seems the present pattern of monsoons only started some ten million years ago, and a theory has been put forward that they have been directly influenced by the vertical uplift of land masses (as a result, one should point out, of horizontal movement elsewhere). In the last million years alone the Himalayas have risen over two kilometres and it now seems likely that winds and precipitation have been directly influenced by this uplift, much as the construction of a groyne or breakwater can lead to the silting up or scouring of an adjoining bay.

'*There rolls the deep where grew the tree. . . .*' The last Ice Age locked up enormous volumes of water during the Pleistocene when what today is known as Dogger Bank in the North Sea emerged as land. It was boggy and forested and became full of men hunting animals with flint weapons, chasing deer and bear and wild ox among the willows and birches. None of this was known until the nineteenth century when widespread trawling started and to their surprise fishermen discovered a lumpy plateau almost the size of Holland lying only 60 feet below the sea's surface. They inferred that this had once been land when they began netting bones and axeheads and moorlog (a kind of peat). The waning of the Ice Age, that era's equivalent of the greenhouse effect, brought an endless close season to the Pleistocene hunt. There must have been a long, mournful period of many centuries as the ice melted and the sea level began to rise again to turn this land between East Anglia and the Netherlands into an archipelago, dozens of scattered islands with heterogeneous collections of hyenas, woolly rhinoceros and mammoth struggling for survival on ever-decreasing patches of territory. Then, at length, nothing but the deep. A mere 50,000 years ago and the forests of Dogger would have been visible from what is now the coast of Lincolnshire. Tennyson, fast in the grip of transience and loss and Charles Lyell's bleak discoveries, had no need of them to complete his vision.

> The hills are shadows, and they flow
> From form to form, and nothing stands;
> They melt like mist, the solid lands,
> Like clouds they shape themselves and go.[1]

---

[1] Tennyson, *In Memoriam*.

II

The Monsters Within

On the coast of a Philippine province there is a small town. On the landward side of the road, set well back among coconut palms and jasmine, is a whitewashed church with a green tin roof, only one of several civic buildings including an elementary school and an abandoned health centre. A legend surrounds this church, one known to every fisherman in town and to every boy who ever jumped off the little coral pier clutching a speargun. The legend *underlies* the church rather than surrounds it, for the story goes that there is a passage leading from the sea to a cave deep beneath it which is the lair of a giant octopus. There certainly is a fissure in the thick cap of fossil coral which covers much of the volcanic basalt of the island's coasts. Its mouth lies about twenty-five feet below the surface at high tide and at night a powerful underwater flashlight shone nervously in reveals no end to its interior.

In the absence of scuba gear there is nothing to be done, since only a madman attempts to explore a submarine cave with a chestful of air. There would be nothing to be done in any case. Whether or not one has been worked on by the legend, this particular depthless black slit does exude a peculiar aura of menace. The water around the mouth is always several degrees colder than elsewhere and very few fish appear ever to venture in. If there really were some great monster lying tucked away inside, a good deal of food would need to swim unwarily in for it to survive. On the other hand, it may be that one way or another much of the town's drains seep into this crack and the creature survives on ordure. It might even have grown so fat it could no longer leave if it wanted to and is bottled up in its coral crypt. The thought of swimming up a sewer to confront a trapped monster is another good reason for not making the attempt.

These are all excuses, of course. The fact is, I am afraid of the place and so is everybody else. From the sunlit surface above its opening one can look across the road through the palms and see the church's corrugated roof. If the fissure really does extend that far, it must be at least 100 metres long. Up in the brilliant daylight the whole legend looks different. The idea of a demon Kraken lying in its lair beneath a church is too naively Filipino, too redolent of Christian mythmaking to be more than the embroidery with which the credulous have ornamented a freak of local geology. Nevertheless, I am not going inside.

Only a mile or two down the coast and not far out to sea the reefs drop suddenly into ultramarine depths. By swimming out over the shallow corals it is possible to pretend one is low flying, hedge-hopping above rough coral terrain, an illusion strengthened by soaring out over this great abyss. So abrupt and powerful is the effect one may even feel one's stomach drop. It was here, over the years, that I would practise seeing how far down I could swim, to set my own private record. Soon I knew every ledge on this cliff face, each downward step of the agonising but exhilarating journey. I knew each level by its peculiar feature—a coral outcrop or eccentric sponge—and also by its ambient light. I knew as well as Beebe where red became grey, where my blood looked dark green in the water. This series of terraces now seems a lunatic perpendicular *via crucis*, every step gained representing pain, but it also stands as a chill measure of ageing. I can no longer reach the very deepest of the shelves I once touched, raising a confirmatory plume of silt still visible from the surface like a triumphant smoke signal impressively far below. I might again, I tell myself, but only if I lost weight since fat is buoyant and means one has to burn up more oxygen to drag it all downwards. Since I can no longer measure my own record I have to estimate it as between eighty-five and ninety-five feet, rather less than the average local teenager can manage when harvesting big white sea cucumbers.

It was here, perhaps to spite myself, I tried 'riding the rock' instead, or using a weight to pull me down. This is how no-limits free-diving records are set. Nobody labours to swim down; they ride the rock and when they let go they hope their still uncollapsed lungs contain enough air for them to claw their way back to the surface.[1] Taken to extremes it is a lethal sport, but I had no intention of going very deep. In the event, the whole business felt faintly embarrassing. There is something foolish about loading rocks into a dinghy, rowing out and jumping overboard clutching them to one's chest. The first was too heavy and took me down so fast I could not 'clear' in time and the pain in my ears made me let go at about thirty feet. The next took me down rather languidly, and it was a pleasure to see ledges I had fought to reach drifting upwards past me like floors in a descending lift. With a subsequent rock I passed my own record and was pulled onwards into unknown territory. I do not believe I ever went further than about 160 feet. There was something disagreeably inexorable about the downward tug. It was not as if one doubted for a moment that one's arms would release the weight before it was too late, but it had something to do with increasing pressure and deepening gloom which I had never experienced without breathing apparatus. Perhaps because the motion was entirely vertical and

[1] No-limits apnea uses a weighted sled rather than a rock for descent, and an air bag for ascent. The current record is held by the Belgian, Patrick Musimu, who in 2005 was pulled down to 209 metres (686 feet) in the Red Sea and survived.

swift one imagined dissolvings like a meteor, leaving a trial of silver bubbles, soon to be worn away to nothing by the rasping caress of the sea.

Beebe had written, 'The only other place comparable to these marvelous nether regions must surely be naked space itself . . . where sunlight has no grip upon the dust and rubbish of planetary air.' The exploration of space and of the deep sea have obvious things in common. Both require venturers to be supplied with complex life-support systems and defended against extreme ambient pressure, whether positive or negative. At a mythic level, however, there are important differences, many of which—in the case of deeps—have to do with the dark.

The famous and fatuous opposition of light and darkness is pre-Socratic in origin, only one pair of many made up of a 'noble' element (right, above, hot, male, dry, etc.) and an 'ignoble' (left, below, cold, female, wet). By the sixth and fifth centuries bc the faculty of vision and the attributes of knowledge had run together in the Greek word *theorein*, meaning both 'to see' and 'to know.' Knowledge was henceforth a register of vision. Ignorance therefore becomes a lack of knowledge predicated on objects not being visible, so darkness equals ignorance. In turn, the dark becomes a source of fear as if a knowledge of visible objects were the only defence against terror and anxiety. By the eighteenth century the light of reason stood for the banishing of primordial fear: literally, enlightenment. Superstition as a concept is a product of eighteenth-century topology.[1] Where the ocean's deeps are concerned several other dualities operate as well, such as up/down, lightness/pressure, outwards/inwards and future/past. To go into space is in some sense to go forwards; to go down into the depths is at a psychic level to regress.

---

[1] This passage originates in a lecture given by Mark Cousins at the Architectural Association in London on 23 November 1990.

Why should this be? Space travel is 'going forwards' in the obvious sense that it involves technological 'progress,' but so does deep sea exploration. It is as though *Homo* viewed himself in spatial rather than temporal terms, as if his history had been one not of eras and dynasties so much as of steady territorial expansion. Maybe the whole of human history might be rewritten, leaving out dates and measuring instead the boundaries pushed outwards by tribes on their way to becoming nations, by earthlings as they stake out their claims to colonise the solar system. Yet even with nations claiming EEZs and seabed rights it never feels an appropriate choice of cliché when journalists call the ocean depths 'the last frontier.' As always, the sea is really less connected with space than with *time*, as if there were a correlation between going deep and going back. Thus the deeper one went the more primitive would be the life forms encountered, the more prehistoric and inchoate.

This must be a comparatively recent idea, post-Darwinian, at any rate, and taking into account a popular version of Victorian scientists' excitement on learning that the deeps were not azoic. The finding of the first coelacanth would have strengthened this, as does every fresh 'sighting' of the Loch Ness Monster. Legends of monsters and sea serpents are at least as ancient as the written word, but presumably it is only after the mid-nineteenth century that they begin to be depicted as prehistoric and corresponding loosely to fossil forms. The Loch Ness Monster is almost invariably spoken of nowadays not as some unknown species of sea snake or eel but as a saurian of prehistoric type. Since this is what people wish to be there, it is faithfully confirmed by all the 'sightings.' It is thus a true remnant of a misapprehension by nineteenth-century science.

Myths of space travel do include visits to worlds at an earlier stage of evolution than our own. Yet even these are often in 'obscure' backwaters of space, as if in the scriptwriters' imaginations space corresponded to a vast ocean in which the most developed regions tended to be those appearing from Earth most brightly lit. ('Rigel Concourse' in Jack Vance's stories is a good example, Rigel being a pure white, first-magnitude star. This is exactly where one would expect to find our outwardly bound pioneering descendants rather than huddled around some dismal cepheid variable out in the galactic sticks.) All this apart, the creatures most commonly associated with space operas as well as with UFOlogy are of an intelligence superior to ours, and with the waning of American paranoia about Communism they tend to be less and less bent on kidnapping and brainwashing. Nowadays space aliens may well incline towards the godlike, beings from whom we might acquire knowledge, enlightenment, light itself, before it is too late.

The mythology of our own planet's oceans is the polar reverse of all this, so much so that the nether world sometimes seems hardly part of the Earth at all. It is worth examining this from the popular standpoint for a moment because it shows how the concept of 'the deeps' relies on a jumble of associative ideas. Far from being likely to find enlightenment the further down we go, then, we expect to meet ever-dumber creatures. Moreover—exactly opposite to actuality—we envisage them near the bottom as still bigger, more terrifying in their mindless strength, and *uglier* . . . in fact, monsters. To this extent they are remarkably similar to the nightmare creatures of the unconscious: tentacular horrors which enwrap and bear their victims down and down to lairs where, in due time, begins the business of the hideous rending beak and saucerlike eyes. The very gradations of sleep itself seem to suggest a vertical descent into annihilating depths, the deepest levels of sleep being those of oblivion. The levels of dreaming, like the layers of the ocean which can support the biggest life forms, lie nearer the surface. In any case, by descending into the sea we would expect to meet the monstrous rather than the divine. Gods are the last things we would imagine finding in the deeps. It is no accident that even the men we encounter tend to be people like Captain Nemo, ominous whichever way we read his name. Astronauts have claimed close encounters with a Supreme Being, but never deep sea divers. Nor should we be surprised. Superior beings are by definition on top, while only the inferior can lurk below. The deeps also remind us of where we suppose we originally came from, what we have left behind. Going back thus to our genetic roots rather than to the sunlit idyll of Eden is a disquieting affair. Did we not abandon our ancestral dark by crawling towards the light?

No; we did not. The sea, to its dwellers, is not a dark place. With exceedingly acute eyes perceiving low levels of light and complex codes of bioluminescence, with sensitivity to sounds, smells and minute pressure differentials far beyond the spectrum of our own senses, it is as pointless to speak in crude human terms of 'light' and 'dark' as it would be when speculating about what a bat sees. A bat 'sees' with its ears with great precision and at speed. In short, there *is* no such thing as darkness. It exists only in the perception of the beholder. Vision, like understanding, does not depend on light.

To these 'oppositions' and their associations (up/down, above/ below, superior/inferior, heaven/hell) should be added striving/ sinking, where the first generally implies upward aspiration and self-betterment and the second is redolent of slummocking on a downward path, of Jack finding his own level while still undrowned. 'Sinking' is also used to describe wretched people glistening with sickness on their deathbeds, as if their problem were only one of weakness and they could no longer resist the force of gravity tugging them down towards their graves. That there might be something subtler at work than these pairs of opposites is suggested by the Latin word *altus* which can mean both high and deep (as it does still in Italian and where *l'alto* means 'the Deep' in an oceanic sense which also lingers in the English phrase 'the high seas'). In the Freudian unconscious, at least, such an idea would not embody a contradiction because there are no contradictions in the unconscious. Entirely antithetical and mutually cancelling propositions can exist simultaneously with not the slightest difficulty.

Perhaps, then, the least strange thing about the Deep is the degree to which it has retained its psychic force, its sonorous and chilling stateliness, its amalgamation of height and depth, of gulfs of space and of time. Almost no matter what is done to the oceans, however much they are explored and exploited, even ravaged and polluted, the Deep surprises us by its resistance to contamination. In this respect its resembles the Moon, which still feels much the same even though we know its dust bears the frivolous prints of cleated boots playing golf. The fact is, it was a different moon on which the astronauts landed, just as it is a different deep which GLORIA deafens with its sonar signals and whose silt is scarred by remotely controlled sleds gathering the sort of things a sled would gather. Neither Beebe nor Piccard nor Ballard ever visited the Deep. They reached various depths, even the ocean bed, but they carried the Deep within them. It is not a space to which there is physical access. Yet an air of mystery, no matter how slight, still surrounds objects retrieved from the depths, even beer bottles and polystyrene cups lowered by the curious. People like to touch things brought up, such as hoppers full of nodules. They like to feel the chill of aeons before it fades, just as they like to handle meteorites and moon rocks. If the ocean vanished tomorrow its mystery would not be found in the sum of its creatures flopping and dying and rotting on its bed. It exists elsewhere altogether, as Tennyson well knew when he capitalised on its high melancholy to express his grief over Arthur Hallam's death, hidden and heightened in a transition: 'From the great deep to the great deep he goes.'[1]

---

[1] Tennyson, *Idylls of the King*, 'The Coming of Arthur,' 1.140.

III

Dating the Earth

William Buckland's *Bridgewater Treatise* maintained that the presence of fossils embedded in sedimentary rocks was definitive proof of Noah's Flood. This work, really a series of lectures, was published in the 1830s, the same decade as Charles Lyell's *Principles of Geology,* whose radical conclusions could scarcely have been more different. By Buckland's day the question of the Earth's age was keenly debated. The natural sciences had evolved to the point where a more serious answer was needed than Mosaic chronology could provide. The supremacy of the biblical version of Earth's creation had already been challenged a century earlier by Newtonians like the Comte de Buffon. Eighteenth-century science had gained enough insight into geological processes to enable it not so much to speculate about how old the Earth was but to wonder how all the vast, slow procedures of erosion and sedimentation and the laying down of fossil beds could have been squeezed into the mere 6,000 or so years allowed by Scripture.

It is possible to argue that until Christianity there was no such thing as time. That is, until after the life of Christ, people's notion of time would have been largely cyclical, based on the regular recurrence of seasonal and astronomical phenomena. Longer periods were simultaneously precise and vague, being measured by carefully preserved familial dynasties such as the lengthy genealogies in the Old Testament. To small rural societies, the Earth was unchanging, as old as legend, as old as their creation myths. It was not possible to apply any external timescale to it because one did not exist. This view of time changed radically with the coming of Christianity. Suddenly, time stopped being a slow, circular continuity and became an 'arrow,' linear, flowing in a single direction from Creation to the Last Judgement. Christ had come as a man in fulfilment of a prophecy, and as a man he had died. He could only do so once, so his life was just an episode—though a momentous one—in a single temporal trajectory which was carrying all mankind with it towards the great Millennium.

Once this idea had taken root, theologians began examining the Judaic scriptures with new motives. They were now less interested in the genealogies as evidence of pedigree than as ways of calculating how many years had elapsed since God had created the world so lately visited by his son. Theophilus of Antioch put Creation at 5529 bc; Julius Africanus at 5500 bc. This was eventually reduced by Martin Luther to 4000 bc and finally Archbishop Ussher produced a date of scholarly precision, 4004 bc, which stood as the problem's definitive solution. This date gained such wide acceptance it was printed in most English bibles with chronologies. It can be found even today in some fundamentalist bibles.

This firm date of 4004 bc had all sorts of consequences for the way in which people thought and looked at the world and soon produced intolerable strains in the burgeoning natural sciences. Even archaeologists began worrying about how societies as complex as that of ancient Egypt could have evolved so quickly. To some it was clear that the oldest pyramids dated from around 3000 bc, only a scant thousand years after the beginning of the World According to Archbishop Ussher. Geology, meanwhile, had progressed to where the publication of Lyell's *Principles of Geology* in 1833 made it clear it was no longer necessary to invoke six-day acts of creation or great-flood catastrophism in order to explain the Earth's structures. All that were needed were the ordinary processes which anyone could observe, plus almost unlimited quantities of time. Once it was assumed that fossil beds could be millions of years old instead of a maximum of 6,000 it created a sort of conceptual breathing space in which many things suddenly began to make sense.

No matter how horrible or absurd their positions may be, there is always an element of poignancy about diehards. It was therefore both comic and pitiful to see a scientist like Philip Gosse confront evidence such as Lyell's and be unable to relinquish his own fundamentalist position. Instead, he went into contortions which effectively destroyed him, bringing down on his fervent and well-meaning head the ridicule even of churchmen. In 1857, only six years after his school zoology textbook, he published *Omphalos* which, as the title implies, had as its centre of gravity the question of the navel. Adam must have had a navel because God had created him as the genotype of the human race. Since Adam had no mother, his navel was surely more exemplary than functional. Similarly, had Adam cut down one of the trees in the Garden of Eden he would have found annual growth rings, although they had been there only days. To deal with this problem Gosse claimed that some things were 'prochronic' or 'pre'-Time. Adam's navel and the tree rings were prochronic. Because the first chicken had been created fully fledged, the putative egg from which it hadn't in fact hatched was equally prochronic. So also were fossils. God, for reasons best known to himself, must deliberately have 'salted' the Earth with fossils in order to make it seem older, perhaps, and maybe even to test the faith of later scientists, especially geologists.

The derision which greeted this pretty notion completely baffled poor Gosse. The Revd Charles Kingsley, whose *The Water Babies* showed he was himself quite capable of imaginative extravagance, was particularly forthright. In his own *Glaucus* he commented bleakly: 'If Scripture can only be vindicated by such an outrage to common sense and fact, then I will give up my Scripture, and stand by common sense.' Gosse was not completely on his own, of course. William Buckland went on thinking that The Flood was a perfectly satisfactory explanation for fossils. But Gosse was a well-known figure, and his flounderings in the name of science made him something of a lightning conductor. Others, too, were disconcerted and depressed by a rationalism which seemed unaesthetic as much as irreligious. Ruskin wrote plaintively in a letter of 1851: 'If only the Geologists would leave me alone, I could do very well, but those dreadful Hammers! I hear the clink of them at the end of every cadence of the Bible verses.'

In the mid-nineteenth century what science badly needed was a reliable temporal yardstick by which to obtain a sensible age for the Earth, and in the following one hundred years many were proposed and tried. Having gone down a coal mine and noticed the higher temperature, Lord Kelvin concluded that it ought to be possible to work backwards using the temperature gradient and extrapolate a date for when the Earth was in its original completely molten state. He suggested an age of 20–40 million years, later putting it at 100 million years. Some geologists considered sedimentation rates, using the annual deposits known as varved layers rather like tree rings. William Beebe thought salt would prove a reliable measure. In *Half Mile Down* he wrote: 'The chemical composition of blood, both in the constituent salts and their proportion in solution, is strangely similar to that of sea water.' The only difference was that our blood is three times less saline than the sea. 'So all we have to do is calculate back and find the time when the ocean was only one-third as salt as the present . . . and we will know [the exact moment of] our marine emancipation.' True, this would not be the age of the Earth itself, but it was still a useful date to fix. Unfortunately it now seems that the ocean's salinity has never varied very much and that prehistoric seawater was not greatly different from modern. This also put paid to the idea for calculating the age of the oceans by assuming they started as fresh water and working out how long it would have taken to leach out that amount of salt from the Earth's crust.

It was only with the discovery that radioactive elements decay at strict rates that a reliable geological or cosmic 'clock' was found. Nowadays the age of the Earth is usually given as 4.7 billion years, but even this is not final. In fact, all cosmological dates are under constant revision and may always be so. It is unclear whether the construct of time's linear 'arrow' can survive quantum imponderables, but in short timescales at least it seems to point to a cheerless future for the human race. The discovery that the universe is not static meant that the Earth cannot last indefinitely, though the precise manner of its end is conjectural. In the 1930s the theory propounded by Sir James Jeans and Sir Arthur Eddington of the 'Heat Death' of the universe was received with the kind of glum consternation that represents less an absolute belief (for few people understood the physics) than a general shift in the recognition of what is plausible. Public moments like these can be looked back on as forming yet another step in a sequence which included the great debates centring around the writings of men such as Gosse, Darwin and Lyell. The trend was inexorable. The human race was slipping out of the hands of God and into a quite other universe determined by the second law of thermodynamics.

Nowadays *Homo*'s fate is commonly thought to be even worse than that, being in his own hands. Death by nuclear destruction, environmental pollution, global warming or in a demographic gridlock of overpopulation are fluently forecast by prophets of one sort or another. Serious minds can be found expending as much effort in predicting man's future span as in trying to date his past. In the face of all this, the choice of a dignified intellectual stance seems limited. The fossil record might be fragmentary, but its message is all too plain. Mass extinction for one reason or another has occurred many times, no doubt more often than we know. It could happen again tomorrow without violating a single natural law. Since the mindless optimism of religion cannot qualify as a serious position, rational man is left in possession of his true intellectual birthright, an exalted stoicism. Bertrand Russell, having read Jeans and Eddington's theory, put it succinctly: 'Only on the firm foundation of unyielding despair can the soul's habitation henceforth safely be built.'[1]

[1] Quoted in John D. Barrow, *The World within the World* (1988).

The sun is definitely lower now. The freezing thermocline in whose upwelling the swimmer was briefly caught has moved on. He is thirsty and weary and finds himself becalmed in an edgy resignation. He has done his purposeful but unrewarded swimming about. Now he wants to preserve his strength for staying alive as long as possible. He is already thinking ahead to another day's floating, taking it for granted that he will survive the night.

For the first time he is considering the possibility of rescue. He has abandoned the idea of finding his boat. He accepts that his directionless first attempts to search for it are more likely to have separated him still further. Even if he did happen to be looking in the right quarter when stern or prow or the tip of an outrigger reared up on a wave there are surely too many intervening waves for anything to be visible now.

He does have a plan of sorts, if that is not too intentional a word for such an impotent state as his. At nightfall, he knows, this area becomes a major local fishing ground. True, many of the boats will have engines over whose unsilenced blatter his shouts may not be heard. But many of the poorer fishermen stop their engines to save fuel and just drift, while the poorest of all will come out here under sail. Since sound travels well over water the swimmer has high hopes that someone will hear.

In the meantime he is once again examining the sunlit depths on the extreme off-chance of rescue from another source. He has heard legends of dolphins helping shipwrecked mariners, of a strange bond which sometimes leads them to aid distressed humans, even occasionally towing them to safety. The sea is empty, however. It seems to him it is a long time since he has even glimpsed a dolphin, several weeks at least. He can remember when it was hardly possible to look at the sea for five minutes in these parts without their breaking the surface, leaping in pairs. Only three or four years ago he would probably have been surrounded by the curious and playful creatures. Now there is nothing. The sea is empty even of their squeaks. The swimmer knows their absence is most likely due to the very fishermen at whose hands he is hoping for deliverance. Why should any remaining dolphin come within a mile of him? Of what use now to invoke 'strange bonds' in so self-interested a fashion when the deal had always been so cruelly one-sided?

6. Fishing and Loss

'From henceforth thou shalt catch men.' Quota-free fishing on the Sea of Galilee.

1.

Modern commercial fishing is a strange, hybrid profession. It affects to be part of the hunting tradition while thinking like a form of agribusiness. It really resembles neither. There is no element of true hunting left in it since the prey is detected by electronic fish finders. At most there is a certain amount of searching to be done in areas where experience suggests it may be worthwhile looking. Nor is it like agriculture, since no farmer ever reaped a harvest without sowing so much as a grain. Only gatherers do that.

There is nothing suggestive of nomadism about the crew of the *Garefowl* which musters at the wharf in Fraserburgh harbour, Scotland, early one morning. It is 2 AM in April with a north easterly wind shivering the lights in the black harbour water. There are six of us. Two have arrived in cars while the rest, like the skipper and myself, have walked out of the warmth of solid stone houses trailing plumes of breath. The town is still. Yet there is a certain alertness about the closed doors and windows, as if they were quite used to opening and shutting behind the menfolk at all hours of day and night, ready to show anxious faces or disgorge rescuers at the firing of a rocket. Donald and I have walked past a huge stack of empty fish boxes, blue plastic crates designed to hold ten or more stones of fish. I thought of Honolulu harbour with its coffins, but the chill black air blowing down from beyond Norway belongs to a planet which does not contain anything as frivolous as tropics or Hawaiian languors. Fraserburgh is a typical fishing community and the *Garefowl* a typical small trawler, high-prowed, blunt, a mere 56 feet long. There is a lot of wood in her. She is quite unlike the high-tech monsters moored elsewhere in the port, as much factories as ships and bristling with arrays of aerials, antennae, DF loops and radomes.

Donald takes us out quite fast through a maze of high stone jetties and harbour walls. Because of the upward-tilting bows there is very little forward vision. Grey, weed-streaked blocks of stone slide past a few feet away, then a white lighthouse, and we are heading for a place where we 'might pick up a few haddock.' This is a good three hours' sailing, but as we may be out for three or more days there is no great hurry. In the intervening time Donald tells me his woes. He sits on a chair with his feet propped on the boss of the wheel. From time to time he glances up at the illuminated compass set flat in the wheelhouse ceiling and twiddles the spokes with one foot. To his left glow two screens: the radar and the brightly coloured display of a Koden fish-finder. Directly before him are a Decca Navigator and Track Plotter.

'Whatever we find,' he says, 'you'll have a ringside seat at a calamity. If we catch a lot, ten to one we'll be discarding most of it as undersized. If we catch nothing it'll only go to show. They're a desert, these waters. A desert. You'll see.'

Conventional wisdom expresses the problem as too many boats chasing too few fish. Indeed, reducing the number of fishing boats in the North Sea, even if it means governments having to buy them out ('decommissioning') is supposed to be one of the main objectives in Brussels. Immensely complex bureaucratic yardsticks are variously applied: tonnages of boat, horsepowers of engine (expressed as kilowatts, however), the numbers of licences issued for boat owners to catch 'pressure stocks' (the most threatened species). These aside, the mass of regulations governing every other aspect of fishing is a bureaucrat's dream. Those prescribing net mesh sizes draw bleak snorts from Donald, while a new mandatory eight-day tie-up period causes an outburst.

'What the hell is a small trawlerman like myself supposed to do? Last month I was at sea five days. How can I make a living like that? I won't work weekends, why should I? My father told me on his deathbed that if ever I worked on Sunday I would have nothing but grief and die poor, and I believe him. So that's eight days gone out of the month. Then these new EU regulations mean I have to go down to the Fisheries Officer and give him twelve hours' notice of when I want to take my tie-up. That means your boat is tied up in the harbour for eight consecutive days, excluding weekends. How can any man reckon in advance when to take his eight days? It depends on the weather, doesn't it? Or repairs. So as it happened the weather was awesome as soon as my tie-up ended. Out of the sixteen days I *might* have fished last month, I could only get out on five. Probably we'd all of us accept eight days off a month, even ten, provided we could pick and choose and didn't have to take them all in a block. It's killing the industry and driving us to crime.'

I imagine bootleg whisky or drug running and he is angry that I am taking what he says too lightly. He thrusts a recent copy of *Fishing News* at me, then snatches it back at once and begins to read from it by the light of the various screens.

'And this is by the editor himself, mind,' he says. ' "It's impossible for any skipper with a boat over eighty feet to stay viable now without breaking at least one of the three main rules—misreporting, using a smaller mesh or making illegal landings. Impossible on the quotas we have now. We're being forced to be criminals."[1] Crazy, you see. Any night you can go down the port in Fraserburgh and Peterhead and watch boats

[1] Tim Oliver, *Fishing News*, 26 April 1991.

illegally landing catches way over the quota straight into lor-
ries. You can wager there's never an FO there. I don't know
where they all go to. They just vanish at the right moment.'

Having reached the grounds Donald loses way to shoot the
net. By the light of the stern spotlamps it flows overboard, followed by a series of plastic floats the size of skulls. The two
trawl doors are also swung into the sea, heavy iron rectangles
whose drag in the water will keep the mouth of the trawl open.
Soon it is all at sixty fathoms, the speed increases to four knots
and there is nothing to be seen but the twin two cables thrumming the length of the boat from the winch for'ard and out
into the darkness astern.

There will be little to do for the next four hours. Ordinarily
the men would crawl into the wooden hutches down below to
sleep or read, but nobody is tired yet. Dawn is only an hour or
two away. Tea is brewed and the wheelhouse fills with men who
have sailed with each other for years, have known each other
since childhood. Two of them grew up in the same street as the
murderer and their contemporary, Dennis Nilsen.[1] It is like sailing in the company of great auks, an extinct species. They are
rationally angry at the botched and muddled decisions by successive governments and the EU; underneath is a more passive
note of sadness that their livelihood is coming to an end and
with it the long traditions of an ancient community.

'There's still a mess of money to be made out here,' points

---

[1] Nilsen, a distant relative of Virginia Woolf, frequently betrays in his
prose and verse a poetic sensibility. This is clearest when he describes a lonely childhood beside this harsh northern sea. His personality, like Tennyson's,
was marked by its proximity. In his prison cell, shut off from the sound of
gulls and waves, he writes, 'I am always drowning in the sea . . . down among
the dead men, deep down. There is peace in the sea back down to our origins . . . when the last man has taken his last breath the sea will still be
remaining. It washes everything clean. It holds within it forever the boy suspended in its body and the streaming hair and the open eyes' (quoted in
Brian Masters, *Killing for Company*, 1985).

out Graham. 'You've seen all the new housing outside Fraserburgh? All owned by fishermen, if you can call them that. Some are little better than boat drivers. They know nothing about the sea but ten years ago they liked the look of all those grants London and Brussels were handing out like blank cheques, so they ordered up boats costing millions and then fish prices doubled twice over and they got rich. Mind you, they're still paying off for those boats and all that flashy tackle. But they think in far bigger terms than men like us can. With that gear they're landing $75,000 worth of fish at the end of a week's trip, having broken every law in the book. They've rigged the nets so the diamond mesh is squeezed practically shut. They might have put blinders on into the bargain—that's another net covering the first. Or if they'd had a governor on their engines to de-rate the horsepower they'll have broken the seal half an hour outside Fraserburgh. No problem. They know how to put a lead seal back on so the fisheries inspectors can never tell. Then they'll probably land their catch illegally. It's a joke, ken? And if they're caught, what's a $7,500 maximum fine to them? It would put us out of business, but not them. Some of those families run three or four boats.'

'We're not saints,' Donald puts in. 'Don't think we never bend the rules ourselves now and then. We have to, otherwise we'd starve. There's nothing left out here for us small folk. It's all been swept clean. Winnie Ewing was right: the stupidest thing we ever did was give up our 200-mile limit around Scotland. Now we've got every Tom, Dick and Harry hoovering up fish as if there's no tomorrow. Which there won't be. We've already got non-EU members out here.'

'It's not only the numbers,' says Graham, 'it's the technology. Progress is killing the fishing industry. Nowadays you can spot a sardine at twenty miles, shoot your net to almost any depth, sweep the sea bare. And nobody has the will to stop it. Take mesh size as a single example. We're restricted to a 90-millimetre mesh. That's fine. We've tried all sorts of sizes and combinations in the past and we've proved you can keep fish stocks up if you use a 90-mill. diamond with an 80-mill. square panel. Imagine your trawl net, right? Like a great sock. It's all 90-mill. diamond except for a strip around the top of the ankle. That's 80-mill. square. Square mesh doesn't close up when you put a tension on it. Now, when your fish see the headline going overhead they're already in the mouth of the net. Their reaction is to swim upwards. If there's a square-mesh panel the small ones escape. They swim right through it while the bigger ones get swept on down to the codend. It's called a codend but it's got nothing to do with cod, ken? It means a sort of bag.'

'As in codpiece?'

Graham says he doesn't know about that. 'Anyhow, we have to use 90-mill. diamond although they're now talking about putting it up to 110, which will catch bugger all. If they do that we're out of business overnight. They're always talking about conservation. Conservation this and conservation that. Well, we've proved you can fish with 90-mill. and still have conservation. So guess what they've just told the prawners they can use? *Seventy* mill. and *two* nets per boat.'

'Crazy, isn't it?' Donald asks the rev. counter. The Kelvin diesel below vibrates reassuringly and jars the surface of his tea into a shimmer of concentric rings. 'It's the truth what Graham's saying. Two 70-mill. nets. Of course you need a smaller mesh to catch prawns, but since they're allowed to land a percentage of fish together with the prawns they just shoot their nets anywhere and take up every last tiddler. So much for conservation. And how is it conservation to allow the Danes to trawl for sand eels off Lerwick? They're only ground up for fish meal and animal food. Thousands of tons of them, just to feed dogs and throw on the fields. You're not going to tell me it doesn't have an effect. There are half the seabirds here compared with a few years back. They lived on sand eels, you ken. It's down to greed, simple as that. Short-term profits today and hang tomorrow. What we've got up here is an entire industry in a mad scramble to cut its own throat.'

A new sun below the horizon is beginning to disclose a haggard sea. As the dawn light strengthens, the surface takes on rumplings like a sheet of thin metal being shaken soundlessly. Out of pinkish, opalescent air the first fulmars arrive as if they knew the *Garefowl* would soon be hauling up. Shortly afterwards she is hove to and wallowing in the choppy swell. The winch growls and wet hawsers begin sighing through sheaves, spraying drops of water. After some minutes the skulls bob up far astern and the cloud of fulmars circles and lands, coming ever closer. The heavy doors come up and are secured outboard, one on either side of the stern. They are freezing cold to the touch, black except along their bottom edges where they have been freshly scratched and polished by dragging along the seabed for four hours. Their cracks and joins are caulked with mud.

Hauling up is an immemorial business. The nets and skills and patience have all been deployed and now it is time to find out how well one's family will live for the next few days. Countless representations of the Sea of Galilee underwrite this moment: images of straining arms and bulging nets, of the glittering harvest of the deep. Nowadays powerful winches do most of the work, but there is still a fair amount of manhandling, of lumps of machinery with great inertia swinging dangerously while men dodge around the edge of the stern above a frigid sea. When the codend is finally hauled out, the picture is suddenly not at all immemorial but dismally contemporary. A soft sack of fish appears, about the size of a bundle of hotel laundry and festooned with rubbish. The heads of flatfish stick out at all angles, eyes bulging with pressure, and from them and between them hang rags of plastic, bin liners, torn freezer packs, lengths of electric cable, flattened orange juice containers. The brutal meeting of still-flapping bodies with machinery is contemporary, too. As the codend is hauled free of the water the fish caught further up the net are already being minced as they are dragged over the hydraulic pulley high above the stern. Their shredded bodies drop into the sea to be pounced on by gannets.

The full squalor of the net's contents is not revealed until the codend is swung inboard and emptied into a wooden sty. Into this mass of bodies smeared with grey North Sea silt wade men in wellingtons, crunching and kicking among the dying, sorting out the unwanted with their feet: paint tins, a length of rusty chain, a battered steel drum, a work gauntlet, two beer bottles encrusted with growths, lumps of torn starfish, clods of jelly, the silvery sack of vacuum-packed coffee with the name of a Hamburg supermarket still legible. The rubbish of a thousand fishing boats and oil rigs and supply vessels is daily fished up, winnowed out and thrown straight back into the sea, building up on the bottom into an ever more concentrated and handpicked stratum of garbage.

There is about the architecture and layout of certain new housing estates on the edges of provincial towns something which makes it easy to guess they have been built on landfill over what, until a year or two ago, was the municipal tip. A clutch of abandoned gravel pits has been steadily filled with a million soup tins, dead refrigerators, burst sofas, outdated phone books and skiploads of dustbin contents. For years it was picked over by flocks of seagulls following a lone bulldozer as it levelled the heaps of rubbish. Finally, it was tamped down and topped off with a layer of soil from an excavation elsewhere in the county, and streets and houses and lawns were planted over the landscaped charnel. One imagines the street lamps as long hollow spikes driven deep into the festering seam to flare off the methane. It is a familiar inland prospect at the blurred borderland between the suburban and the subrural where the terrain is anything but pristine.

Far off the coast of north-east Scotland at dawn, however, the view appears untouched. The blank light lifts like an ancient eyelid disclosing the glint of curved ocean in its time-less gaze. Gannets furl themselves and dive hard into the black-green surface; they are visible many feet below, stream-lined as tenpins. In the cold, taintless air there is this same arc-tic clarity and there comes the memory of last night's weak flames and tattered banners draped across the northern heav-ens, the flickerings of an aborted aurora. To look into this grand expanse of movement and steely colour is still to see what the Vikings saw, maybe, only for a moment, the idea of North; a vision which bursts as the eye catches on the horizon a structure hardening in the light—part Eiffel Tower and part Christmas tree. This single glimpse of an oil rig pricks and deflates everything. All too obviously it is no ship, not floating but *standing*. In the instant of this perception the North Sea sinks as though a plug has been pulled. It is no profound, pure abyss after all but a lake shallow enough for man-made girders to stand on the bottom and poke through. Then also comes a memory of something Graham has said, that to trawl along the various pipelines for the different species which favour them as habitats (oil pipes are warm, gas pipes cold) has the disadvan-tage of filling the net with a mass of junk and hardware dis-carded by the men who laid the pipes and who regularly main-tain them. Their courses are marked by tons of assorted soft-drink cans, plastic sacks of jettisoned nuts and bolts, paint drums still half full of toxic sludge, hundreds of fathoms of steel hawsers, sunken crisp packets.

One night (Donald had interrupted) his brother-in-law out of Peterhead was hauling up and when the codend swung into the arc lights above the counter, every man aboard froze where he stood. Crushed into the meshes was the face of a girl looking out at them, her mouth open in a yell, her eyes wide. Partially lost among the plaice and whiting and dogfish were her twisted limbs. When the catch was released nobody wanted to wade into the bin to dig her out from beneath the bottles and squid and halibut, not least because there was movement everywhere as if things were trying to struggle up from the bottom of the heap. Eventually some brave soul pulled her out from beneath a heaving monkfish: a torn and deflated life-sized sex doll. Inside her mouth, moulded into a red-rimmed O of insatiable accommodation, were hermit crabs. She, too, went back over the side, twentieth-century mermaid, Jenny Haniver herself, probably modelled from the by-products of the very same North Sea oil her roustabout lover had been helping to extract.

Now, on the shelter deck whose hatch is closed against the wind, Graham in wellingtons crunches and kicks the dying, scooping up fish a stone or two at a time on to a sorting tray where Ian and Gordon quickly work through the catch. Into one plastic box they throw angler fish, into another cod, into a third plaice and sole. Then with a sweep the tray is cleared and down a stainless-steel gutter leading out over the scuppers shoot all the unwanted creatures together with bottles, plastic, bits of wire and shells. Fish of all kinds, wriggling and still, a hundred meals chucked back over the side in a slurry. Outside the gulls dive but pick and choose. The gannets prefer whole fish while the fulmars are really waiting for the gutting to begin. Behind the *Garefowl* stretches a sinking slick of bodies, for even the ones still wriggling will die. The violence of their capture and the hauling up through 60 fathoms has ruptured capillaries, made flotation sacs bulge out of mouths. Those fish with patches of scales torn off will anyway become prey to disease, to worms and saprophytes. One way and another they all drift back down to the lifting, swaying layer of refuse.

Any indignation at such 'discards' is not the conventional 'Think how many Third World families they would feed,' since the same can be said at any other point in the chain of Western food production and consumption, from the processing of frozen vegetables to expense account luncheons. The instinctive rebellion is more against the blithe inefficiency of a system whose regulations make it mandatory to junk a great percentage of what it catches; regulations, moreover, designed expressly to protect immature fish and replenish stocks. How powerful is this reversal of a famous miracle. Once, two small fishes allegedly fed 5,000 people. Our achievement is to make 5,000 fish feed two.

It is utterly violent, this daily wrenching of indiscriminate tons of living creatures up into the air, picking out those which happen to meet current laws and requirements and returning tons of unwanted corpses to the sea floor. The sea floor itself has meanwhile been scoured by the trawl doors, its ecology (already labouring beneath a top-dressing of litter) further battered by the wholesale disruption of colonies of species that have nothing to do with human gastronomy and everything to do with the marine food chain: starfish, crustacea, weeds, corallines, urchins, jellyfish, algae. In addition there are the eggs of fish species such as the herring which are laid demersally, adhering to the seabed. It is never directly seen, this massive damage caused by the corridors scraped across the floor of the sea twenty-four hours a day, every day of the year, all over what were once the richest fishing grounds in Europe. (Silent beneath their dank quilt of freezer bags lie the lost birch forests of far-off Dogger.) What is seen in the codends of the nets is a representative sample, as the rubble in a skip is the microcosm of a building site. I wander about the *Garefowl*'s stern, picking up dabs and other flatfish which have fallen to the deck and lie like dead leaves behind stanchions and around the bases of hydraulic gear, throwing them to the gannets.

Maybe the only way to justify these creatures would be to be forced to eat them all, starfish included.

Over the next couple of days the boat sails back and forth, shooting and hauling up, shooting and hauling up. Twice the net comes up torn after a menacing concussion has thrummed through the boat. Each time Donald has hurriedly shut off the engine, for it means the trawl has hit a rock or some other obstacle. The North Sea is littered with wrecks. While Donald has marked many of them on his rolls of Decca Navigator traces there are always unmarked ones or fresh hazards.

'Bastards,' says Ian as the trawl comes up with the coils of a hawser snagging the headline. 'That's a towing cable from another trawler. Not lost, ken, just dumped. You've got two wire towing cables, right? About three or four thousand quid the pair of them. On average they'll last ten months. There's no scrap market for the old ones. You're forced to dump them. We take ours to a place where there's a hard bottom, preferably near a wreck, so nobody will be trawling there. But if you're not a local and don't aim to come back you don't care where you chuck your cables, do you? Bastards.'

Although the catch is poor—Donald says it will not cover the costs of this trip, let alone pay anyone aboard—spirits in the wheelhouse are not low. Rather, everybody is philosophical. Sometimes you do all right, sometimes not. That's fishing. And what can you expect when the fishery is at its last gasp? (The radar screen is bright with the dots of other vessels: two Danish industrial ships, a German beamer, two Belgian trawlers, a Frenchman and a Dutchman.) Instead, conversation turns to what bad seamen some of the younger folk are, how they hardly know anything about the sea, but technology helps cover up their ignorance. The automatic pilot makes navigation simple, everyone agrees, but it is also dangerous since it lulls the inexperienced into not paying enough attention and is often the cause of collisions.

'My father would have died of shame at the thought of colliding with another boat. Nowadays it's a bit of a joke. Insurance, you see.' Suddenly the voice of an old, tight Scottish community is heard as they reminisce about the conventions, taboos and superstitions which even today rule some fishermen's lives. 'My father was ashamed to do plenty of things people do now without a thought. Other things he just *couldn't* do, like put to sea on a Sunday. And there were certain words you wouldn't mention in front of him, he was that superstitious. You wouldn't dare say "salmon", for instance; it brought awesome bad luck. You had to get round it, say "the silver fish" or something like that. Nor "minister". Nor "pig".'

'Nor "rabbit", come to that,' puts in Graham. 'It was a "four-footer". And nobody ever wore green. As for bodies, they'd never touch a floater. Never.'

Their superstitions must have put them in the strangest positions. Men who would not touch drowned bodies would man the lifeboat and Donald's father, who had been coxswain, could not swim.

'You mean he was a lifeboatman and couldn't *swim*?'

'That's right. No more than I can, and I've been coxswain myself.'

It turns out that only one of the *Garefowl*'s crew can swim. Donald was once knocked overboard wearing thigh boots, which promptly filled and prevented him from kicking himself afloat. He managed to catch a rope in time. 'It wasn't good,' is his description of being pulled below. 'By rights I shouldn't be talking to you.'

It is hard to know what to make of all this. They assert that at most ten per cent of Fraserburgh's older fishermen can swim but a few more of the younger can. They say this is because the young ones have been on survival courses run by the oil companies, but a more likely reason would seem that nowadays schools take their pupils to heated swimming pools. Undoubtedly it is evidence of a fishing community's strange and fatalistic relationship with the sea. In Britain, as elsewhere, there is a long tradition of neither fishermen nor sailors being able to swim. This is disdainfully brave or plain damn silly; else it stems from a dark belief that once you have fallen into the sea's grasp you are done for *by rights*. Or did a Puritan dislike of undressing, together with these latitudes' naturally frigid waters, always make of the sea a medium which was quite properly fatal?[1]

We turn for Fraserburgh with fourteen boxes of fish. We must have discarded at least the same amount, more if dropouts are included (the fish that are shed from the net as it is hauled up). Donald thinks that if instead of throwing away undersized fish trawlermen were obliged by law to box and land them, the extra trouble would be enough to force them to adopt larger mesh sizes. Privately, I think it a pity he and his mates are so conservative in their tastes they will not even keep something back for themselves. They could eat fresh fish at least once a day (in three days we have eaten fish for one meal out of a total of nine), and I try to convince them how delicious most of their rejects are. In vain. No amount of insistence will make them taste octopus, crab, squid, gurnard or several other species. Donald says, 'You get about twelve stones of octopus

[1] To redress any implication that it is nowadays only in places such as isolated Scots fishing communities that people's lives are still affected by superstition, it is worth remembering that at the end of April 1991 the nation's Driver and Vehicle Licensing Centre in Swansea decided to stop using the number 666 on registrations because motorists' belief in the figure's Satanic connotations was leading to a significantly higher accident rate. According to a DVLC spokesman quoted in the London *Evening Standard* (7 May 1991), 'you see the number 666 in front of you and it makes you feel nervous. And, because you feel nervous, you bump into him.' Compare this with the episode of the house President Reagan and his wife Nancy bought in Bel Air for their retirement. Its address was originally 666, St. Cloud Road. They insisted the number be changed to 668. Of an American president's day-to-day movements and decisions being determined to a large extent by the advice of professional astrologers we say nothing, except to add that the French socialist president, François Mitterrand, was similarly guided.

to a box. On average they'll fetch four or five quid a box, maximum ten. Ten quid for nearly 80 kilos of fish? Isn't worth the effort. Dump them . . . *Eat* them? I'll never eat one of those things. They all go to France. Places like that.'

Finally, they hardly seem like people gathering food at all; it is merely something that can be sold. It might be manganese nodules we are bringing in. They are heading home to plates of Aberdeen Angus in thick gravy with tatties and bashed neeps.

There are several very curious things about the present state of the North Sea fisheries, not the least being that it is nothing new, not even the platitudes so often repeated in small wheelhouses: 'The fisherman is his own worst enemy' and 'Progress is killing the industry.' The second half of the nineteenth century saw mechanisation applied to fishing, advances which brought steam power to boats and winches and helped preserve and distribute catches by means of refrigeration and the railways. As early as 1863 a Royal Commission enquired into whether the spread of trawling was perhaps begin- ning to deplete the North Sea stocks and decided it was not. By the turn of the century there was no longer much doubt on the part of anyone involved, with the exception of the government itself. Several peers knew better, but they were sportsmen. Lord Onslow, who had studied the matter at first hand and had already noted with distaste that the trawl crushed fish to death while the driftnet strangled them, determined to push legislation through parliament which would at least restrain trawlers from wasteful fishing. He proposed:

1. closing certain areas, either on a seasonal basis or for some years at a stretch;
2. extending the three-mile territorial limit;
3. a compulsory increase in net mesh size;
4. prohibiting both the landing and sale of flat fish below a certain size.

*

His attempt failed. Meanwhile, a fisheries expert named Aflalo was documenting the destructiveness of North Sea trawlers.

It appears that one shrimp-trawl, working in the Mersey, took 10,000 baby plaice in a single haul, while another in the same district brought up over 250 small soles, 900 tiny dabs, nearly 300 unmarketable whiting, 18 little skate, and some hundreds of useless plaice. When to this is added the fact that all this waste fish was accompanied by only twenty quarts of shrimps, some estimate may be formed of the terrible destruction for which such agency is responsible.[1]

That was written nearly ninety years ago. He went on to put the entire problem in a way which could scarcely be bettered today. Having referred to North Sea practices as a 'conspiracy of depletion,' he recommends fish farming as 'making it possible to sow as well as to reap' and concludes:

It is when man shall have discovered the means of restocking the sea and of controlling its supplies that his 'dominion over the fish' will be perfect. The power to deplete, which so far marks the utmost limit of his advance, is mere tyranny. Dominion should embrace a more benevolent sway, and to that end no doubt the efforts of science and the might of law will presently join forces. It is to be hoped that the present friendly collaboration of the Northern Powers in the great sea in which they have a common interest may be the basis of a lasting harmony, more durable than any evolved in Utopian deliberations at the Hague.[2]

[1] F. G. Aflalo, *The Sea-Fishing Industry of England and Wales* (1904), p. 56.
[2] *Ibid.*, p. 375.

A century on, hardly a word of this needs to be changed except to substitute 'Brussels' for 'the Hague' and to note that fish farming is an established industry, but in areas such as fjords and estuaries annexed for this purpose. There is no attempt to restock free-for-all fishing grounds like those of the North Sea. Much of the problem is that science has not yet lived up to Aflalo's expectations and extraordinarily little is still known about the behaviour of fish, of their breeding and feeding and migratory habits, of the influence on them of changes in current, temperature and plankton levels. As Donald sardonically remarked, 'The government have no idea when cod and haddock spawn.'

In 1902, two years before Aflalo made the above remarks, the International Council for the Exploration of the Sea was founded. It is this same organisation, ICES, which today is responsible for making the scientific assessments of fish stocks on which the EU bases its quotas: the so-called TAC or Total Allowable Catch. (It should be noted that TACs apply to fish *landed*, not to fish *caught*. It is not illegal to catch and kill large numbers of immature fish.) From the year of its founding ICES became a focus for all aspects of biological oceanography and fisheries research. Put succinctly its remit, according to one ICES member, was 'to tell the commercial world how far greed might safely go'.[1]

ICES's first task was to discover why fish have a habit of coming back to the same grounds for decades at a stretch, long enough to generate a local fishing industry, and then one year vanish altogether. It was supposed that a change in the currents carrying plankton might be one explanation. Other ICES scientists tagged fish in order to study their movements. Still others concentrated on refining a method of dating fish which had originally been used to determine the ages of carp, that of

[1] D'Arcy W. Thompson, 'The Voyages of the *Discovery*,' *Nature*, 140 (1937), p. 530.

counting the growth bands on a single scale. They noted that fish grew faster in summer and hence the bands on their scales became wider. It was a Norwegian, Johan Hjort, who finally discovered that the dynamics of fish populations are quite unlike those of terrestrial animals. Any single haul might well contain fish from two to twenty years old, and in very uneven distributions. He invented a terminology, calling fish born in a particular year a 'year class,' then set about trying to explain why there were good and bad year classes. A good year class would dominate catches for several years at a stretch while members of a bad year class, even though they might have grown well as individuals, were thinly represented and for shorter periods. This was mystifying. Some unknown factor—not obviously the weather—evidently allowed great numbers of fish to survive one year but then killed off young 'recruits' in successive years. (No less mysterious was Hjort's describing fish in terms of a military academy. This had no connection with 'schools' of fish, either, which derives from the Dutch word for shoal.)

On the other side of the Atlantic, meanwhile, ICES had been anticipated by all of thirty years by the US Fish Commission. This was the brainchild of Spencer Fullerton Baird, who set up his base in a tiny village near Cape Cod named Woods Hole, which he knew and liked from having spent his holidays there. His laboratory, together with its research ship named the *Albatross*, offered the best facilities in America for marine biology. Baird died, much lamented, in 1886, not before T. H. Huxley had praised him for his work on fisheries which had yet to be done in British waters:

> If the people of Great Britain are going to deal seriously with the sea fisheries, . . . unless they put into the organisation of the fisheries the energy, the ingenuity, the scientific knowledge and the professional skill which characterises my friend Professor Baird . . . their efforts are not likely to come to much good. I do not think

that any nation at the present time has comprehended the question of dealing with fish in so thorough, excellent and scientific a spirit as the United States.[1]

Two basic questions which Hjort posed have still not been answered to complete satisfaction. One was: Which period of a hatchling's life is the most critical? and the other: What are the factors which most affect its health and development? It is now believed that a good year class results if there is a rich supply of plankton during the first ten days after the larvae hatch. In view of the advances over the past one hundred years in science of all kinds, including biology itself, this tentative conclusion scarcely shows a great breakthrough in knowledge of the life cycle of fishes. This in turn makes it odder still that governments on both sides of the North Sea have done so little to curb the 'conspiracy of depletion.' After all, there has been overwhelming evidence of serious over-fishing throughout this period, among the most convincing of which was the effect of World War I. Submarine warfare and minefields closed many fishing grounds for four years, allowing fish stocks to regenerate. With the end of hostilities in 1918 North Sea fishermen landed fabulous catches, not simply greater numbers of fish

[1] Quoted in G. Brown Goode, *The Smithsonian Institution* (1897). To judge from the way he was remembered years later, Baird was an exceptional man. In 1918 a friend, Edwin Linton, paid tribute to him in elegiac vein:

I remember the day and the hour. It was afternoon, and the tide was low. I recall a picture of a red sun hanging over Long Neck and reflected in the still waters of Great Harbor, of sodden masses of seaweed on the dripping piles and on the boulder-strewn shore; and there rises again the thought that kept recurring then, that the sea is very ancient, that it ebbed and flowed before man appeared on the planet, and will ebb and flow after he and his works have disappeared; and in a singular, indefinite impression, as if something had passed that was, in some fashion, great, mysterious and ancient, like the sea itself.

(From E. Linton, 'The Man of Science and the Public,' *Science*, 48 (1918), p. 33.)

but bigger specimens. This bountiful state of affairs lasted a short while, then deteriorated again so that by the mid- to late 1920s fish stocks were back down to where they had been before 1914. The obviousness of such evidence is never enough to convince governmental exponents of *laissez-faire* policies. Even in the middle of World War II, Dr. E. S. Russell, the Director of Fishery Investigations at the Ministry of Agriculture and Fisheries, could in the same breath urge against over-fishing during the post-war reconstruction and also think it 'probable that surface fish, like herring and mackerel . . . although a great source of food and wealth, exist in stocks too large to be depleted by fishing'.[1]

To put it bluntly, nobody has the remotest idea of the full long-term effects of continuous over-fishing. It has recently been suggested that it might be causing genetic changes in fish stocks as the fish, fighting for survival, are pressured into earlier sexual maturation at the expense of growth.[2] Shore stocks may also be driven wholesale into deeper waters, as has happened in the Grand Banks fishing grounds off Newfoundland, with unknown consequences for the area's ecosystem. Least of all can anybody say what impact the powerful krill fishing industry will have on South Atlantic waters, on baleen whales and penguins or—come to that—the entire food chain of Antarctica. By the mid-1980s annual catches were peaking at half a million tonnes. Since then, they have averaged 120,000 tonnes. It remains to be seen whether the fishery's regulatory body, the Convention on the Conservation of Antarctic Marine Living Resources (CCAMLR) can impose precautionary catch limits before this valuable fishery expands further.

---

[1] *The Times*, 10 March 1942.
[2] See Richard Law, 'Fishing in Evolutionary Waters,' *New Scientist*, 129, no. 1758 (2 March 1991).

One day in January 1988 two companions and I were in a skiff midway between two islands in the South China Sea when we spotted something curious in the water ahead. There was a ragged patch like floating seaweed or the top of a reef as the tide begins to fall. We knew the area well; we were far out in a deep channel where there were no reefs. Also puzzling were the antics of a small white tern which was standing, fluttering and fluttering its wings as if in some difficult feat of balancing on this patch. As we neared we could smell it before we finally identified it as a mass of rotting animals. We identified fish, a baby dolphin and various bird carcasses. The tern's desperate agitations increased with our approach and we could now see that its feet were caught in the nearly invisible meshes of a ghost net, a fragment of drift net which floats and continues to catch almost any animal that comes into contact with it. With some difficulty we freed the little tern, the bottom of the skiff whispering over nylon and fins and bumping softly among the heavier corpses. There was nothing else to be done. Sooner or later, the gases of putrefaction having been released, the mass would sink beneath its own weight until fresh gas was generated or its heavier contents broke up sufficiently to filter out between the meshes. Then it would rise to the surface once more.

After Japan first began using nylon monofilament nets for drift netting in its own coastal waters in 1976 there was a great expansion of pelagic drift netting. For years the Central and North Pacific areas were the principal focus of public and government attention, tending to obscure the fact that the practice was widespread in coastal areas as well. In archipelagic seas it caused immense damage to marine and bird life as well as affecting the livelihoods of local fishing communities. Pelagic fish are those which feed at or near the surface, usually in large, dense shoals. To catch them, drift-netters hung an invisible curtain of nylon mesh forty feet high in the sea and left it to drift with the tides, currents and winds. Drift netting was a long-established technique, one that had been used for over a century in British inshore waters. What was new about the practice since 1976 was the use of invisible and virtually indestructible nylon mesh.

For six months of the year the North Pacific was invaded by more than 1,500 ships, mainly from Japan, Taiwan and South Korea, together forming the world's largest fishing fleet. Every night each ship laid between thirty and fifty miles of net, leaving it to drift for six to eight hours, so that in these latitudes on any night there could be anything up to 50,000 miles of net deployed, about enough to encircle the planet twice. The ships were highly efficient, with the latest electronic means for detecting shoals, especially those animals rising vertically each night as the Deep Scattering Layer, and some with additional sonic aids for herding them more tightly. Such nets catch almost anything which touches them, including birds and especially cetaceans, to whose sonar the filaments are invisible. Marketable fish were kept; non-target species and those of no immediate economic interest were dumped overboard by the ton. It is a system which has been trenchantly described as 'stripmining the oceans'.[1] From the fishermen's point of view it is cheap and simple; from that of the ecosystem the effects cannot be determined since so little is known. But its results are not invisible. By 1990 even Japanese drift-netters were admitting they saw far fewer dolphins, seals, juvenile hump-back whales and seabirds than only five years earlier. Japan had long since banned drift netting in its own territorial waters because of the threat of ecological collapse.

In a landmark victory for conservation, the UN General Assembly passed a resolution banning drift netting on the high seas from January 1993, though it was still permitted in coastal

---

[1] This vivid phrase was coined by the Hawaii-based Earthtrust group who painstakingly brought to light much of the early information about this secretive industry.

waters. The EU banned its fleets from drift netting in the Atlantic and the Mediterranean from January 2002, although after strenuous lobbying by Denmark, Sweden and Finland the Baltic was exempted from this protection. The practice also continued in other limited areas where exemptions were obtained, such as the UK coast between North Yorkshire and the Scottish border, where from June 2003 just sixteen remaining salmon and sea trout netsmen were allowed licences. But the trend is obvious: the practice is at last on the way out. To be sure, illegal drift netting still continues the world over, but it is steadily being reduced by better aerial spotting, reporting by ships and general enforcement. But rising seafood prices allow no grounds for complacency, and constant vigilance will be required.

The activities of the drift-netting industry first came to urgent attention around 1983, and then only because Japan had formally requested US permission to go drift-netting off Alaska within the 200-mile EEZ. The intention was to intercept Pacific salmon on the way to their river spawning grounds. The US granted the permit, believing that these stocks were heading for Russian rather than North American rivers. Later, when the Japanese tried to renew it, the questionnaire they had to submit unwisely revealed that a typical season's catch included an estimated 14,000 Dall porpoise as well as three-quarters of a million seabirds. (At the time the Japanese salmon fleet comprised 172 vessels, each with nine miles of net, totalling a mere 1,548 miles as compared with that currently being deployed by the North Pacific red squid and tuna fleets.) As it happened, the salmon stocks were North American after all. This incident provided an excellent demonstration of how little is known about the movements of fish. The heirs of Spencer Fullerton Baird had misread the migration patterns of the most economically important species in Alaskan waters. This is to leave aside the ethics involved in the assumption that since they were Russian salmon they were fair game for plunder. It was only when the Alaskan and Canadian salmon fishermen began to suffer ruin from the depletion of their stocks that the error was acknowledged. In 1988 Japanese fishing boats were banned from US territorial waters, two years after a similar ban by Australia.

Yet even today, East Asian markets are regularly glutted with salmon still being trawled illegally in Alaskan waters. This fleet probably takes 50,000 tons of salmon annually, and Earthtrust has estimated that the number of 'dropouts' and injured and dying 'escapees' equals the same amount again. If this can happen within the United States' EEZ, it is not hard to imagine what goes on in international waters. Policing is slowly improving, and there is no question that the steady outlawing of drift nets has struck an important blow on behalf of wildlife. But those with a sense of history will retain their pessimism as a thwarted billion-dollar industry develops instead the technology for precisely targeting species, with the danger of their eradication (as in the case of the majestic bluefin tuna). 'Quotas' will once more be fixed by authoritative-sounding international bodies who cannot tell how much loss the ocean has already borne, nor guess what it still can bear. If the North Sea, surrounded as it is by a community of developed nations with access to the best scientific information, can be systematically ruined through conflicts of interest and political expediency, what real hope is there for the Pacific? Who is to police its vast, unclaimed areas, and to enforce what law? Under the virtuous flag of an abuse redressed, cynical attrition is as likely as ever to slither around laws and evade patrols, just as it does in the tiny pond of the North Sea. The issues will float up and down for years to come like ghost nets with their rotting cargoes.

For now, even people with no immediate connection with fishing have noticed a decline in Pacific wildlife. As a geophysicist aboard *Farnella* said one afternoon, surveying from the rail the empty expanse of ocean on which floated a sheet of tar-stained plastic where eight or even five years before a school of dolphins might have sported, 'We don't deserve this world.' The drifting refuse, the absence of any sign of life for two weeks but for a few flying fish and the occasional mournful bird was bound to make anyone take stock. '*Homo sapiens sapiens*. In this single century we've slaughtered a thousand times more people than all the Genghis Khans of history put together. Into the bargain we've laid waste our planet. Not bad going for a mere 100 years. And look what we've got in return. Machines for mapping the ocean floor and a brave new race. We've had the Beaker Folk and now we've got the Consumer Folk. Tesco Man. *Homo supermercatus.*'

The trait in the human species of harvesting first and assessing the consequences at leisure is clearly a genetic inheritance. Unlike felines, which eat their fill and walk away, we are in this respect closer to the canines such as foxes, which kill an entire roost of chickens they will neither eat nor bury for later. There is something hopeless in *Homo*'s mixture of brutality and compassion, a cross-purpose of muddlement. Even as Spencer Fullerton Baird was founding Woods Hole, Charles Wyville Thomson was trying to observe the creatures of the North Sea. He wrote of English fishing smacks being welled to supply fresh cod for the London market (this was before refrigeration). A large, square tank was built amidships with holes for fresh seawater to circulate. The fish, he noted, were oddly tame:

> It is curious to see the great creatures moving gracefully about in the tank like goldfish in a glass globe . . . They seem rather to like to be scratched, as they are greatly infested by *caligi* . . . One of the fish had met with some slight injury which spoiled its market, and it made several trips in the well between London and Faëroe and became quite a pet. The sailors said it knew them . . . It was always the first to come to the top for the chance of a crab or a bit

of biscuit, and it rubbed its 'head and shoulders' against my hand quite lovingly.[1]

Treating a codfish as another Englishman might treat his hound may have struck his contemporaries as eccentric, especially since they were accustomed to a reign over the animal kingdom which was in general less than benevolent and often maniacal. Accounts of the huge eighteenth- and early nineteenth-century massacres of creatures such as whales, seals and sea lions make painful reading; even at the time, explorers and others sensed that something was not right. W. H. B. Webster, the surgeon aboard HMS *Chanticleer* during her voyage of 1828–30 in the south Atlantic, wrote:

> The harvest of these seas has been so effectively reaped, that not a single fur-seal was seen by us, during our visit to the South Shetland group; and, although it is but a few years back since countless multitudes covered the shores, the ruthless spirit of barbarism slaughtered young and old alike, so as to destroy the race. Formerly two thousand skins a week could be procured by a vessel; now not a seal is to be seen.[2]

Captain James Weddell, making much the same journey five years earlier, had noted that both sea lions and fur seals were almost extinct on South Georgia, 'not less than 1,200,000 skins' having been sent to the London market, together with 20,000 *tons* of 'sea-elephant' oil. He evidently liked penguins, observing that 'Sir John Narborough has whimsically likened them to "little children standing up in white aprons".'[3]

When fur seals became scarce, the 'sea-elephant' (elephant

---

[1] C. Wyville Thomson, *The Depths of the Sea* (London, 1874), p. 59.

[2] W. H. B. Webster, *Narrative of a Voyage . . .* (London, 1834), Vol. II, p. 302.

[3] James Weddell, *A Voyage towards the South Pole* (London, 1825), p. 5 3 .

seal) was turned to for its oil, and Weddell noticed the animal's pitiful docility. 'It is curious to remark, that the sea-elephant, when lying on the shore, and threatened with death, will often make no effort to escape into the water, but lie still and shed tears, merely raising the head to look at the assailant; and, though very timid, will wait with composure the club or lance which takes its life.'[1] In order to extract the oil from the seals' blubber it was flensed and rendered down in vast iron cauldrons. Fuel being extremely short in these latitudes the sealers burnt penguins, each of which contained about a pint of oil. Finally, the supply of seals ran out and the sealers had to turn to the penguins themselves. 'In common with all Antarctic creatures,' observes Edwin Mickleburgh,

> they had little or no fear of man and it was therefore a simple business for the sealers to drive them into pens where they were clubbed and flung into the bubbling pots. It is recorded that some birds, in the interests of faster production, were driven down crudely constructed gangways directly into the pots where they were boiled alive.[2]

Not that this sort of thing was confined solely to sealers and whalers. W. H. B. Webster, calling at Buenos Aires on his way south, had already heard a local resident describe how he had 'once sold a flock of sheep, amounting to two thousand, at 1s 6d. per head, for the sole purpose of fuel for a brick-kiln.'[3]

Looking once again at the North Sea and the Pacific Ocean 170 years later, we can see that there is little new about the creation of semi-deserts where once teemed marine and bird life. Nor, in its capacity to kill indiscriminately, does modern driftnet fishing really mark a new departure. Probably the only

[1] *Ibid.*, p. 136.
[2] Edwin Mickleburgh, *Beyond the Frozen Sea* (1987), p. 31.
[3] Webster, *op. cit.*, Vol. I, p. 83.

novelty is that of pollution, of animals being sieved from the sea and replaced with bottles and sheets of tarry plastic. Even in this matter the same muddlement and cross-purposes can be found, the same ironies. Thus, scientific bases in Antarctica, set up precisely to study a pristine, nearly sterile environment, have themselves polluted their surroundings. The worst offender has been the US base at McMurdo Sound, where surplus bulldozers have been casually tipped into the sea and open-pit burning and landfilling of wastes were everyday practices for years. Efforts are continuing to clean it up and many another base, newly uncomplacent, is discovering how expensive and inconvenient it is having to ship out hundreds of barrels of frozen faeces and urine in addition to the other refuse it generates. There is an irresistible parallel to be drawn between Antarctica and another pristine—and wholly sterile—environment: space. Since the space age started in 1957, 6,600 spacecraft have gone into orbit, of which some have fallen but plenty remain. In addition to these and their defunct rockets there are an estimated 500,000 fragments of junk, including screwdrivers. If in the next three decades the amount of debris in low Earth orbit increases at the same rate it will total three million tons and render any further human space exploration prohibitively dangerous. Various hugely expensive schemes for cleaning up space are currently being considered.

Until recently, I was finding it easy to compare modern fisheries such as those in the North Sea with what it pleased me to think of as 'true hunting.' Because I spent time each year hunting my own food in the sea I deployed sentimentalities in my own favour. Certainly, when one takes on individual fish with a home-made, elastic-powered speargun of very limited range and accuracy, using only the air in one's lungs, the odds are heavily in the fish's favour. No doubt also it attunes one to things previously obscure. In the lengthy daily process of stalk-ing food one learns much about different species of fish and their behaviour. One also becomes observant of other phenomena, noticing weeds, currents, sudden thermoclines, coralline animals and the local benthos, light and shade and colour. One also discovers things about one's own body: how to control breathing and how to lie at different depths, as well as the graph which plots time spent in the water against increasing frequency of urination, the remarkably dehydrating effects of three hours' constant diving and underwater swimming. There are unlikeable character traits to confront, too, including those pairs of seeming opposites: callousness and fear, impatience and hesitancy.

There is one particular sentimentalism about hunting which all of a sudden I do not like at all, maybe because I once partly subscribed to it myself. It is that which speaks of a deep, quasi-mystical 'understanding' between hunter and prey a sort of mutual respect where after hours of effort the hunter is half pleased when his quarry escapes or, conversely, when it seems almost content to die. Presumably this derives from the humbug of chivalry and the codes of jousting. The thing in hunting is to win. When the novelty of the experience has worn off and the basic technique has been acquired, there remains the task of getting one's food as quickly and efficiently as possible because there are plenty of other things to do such as collecting firewood, making another spear, repairing a rickety fish drier or just sitting under a shady tree. At this level' hunting is simply gathering food, a necessary and often pleasurable chore. It is quite different from those grand, allegorical duels between old men and the sea, or grizzled captains and white whales. Yet to fish day by day off the same stretch of shore and, where there is a long fringing reef, off the same groups of corals, is to see from within the impact of local fishing. Nowadays I prefer to swim out beyond the reef, to go out at night and instead of killing parrotfish in their holes or mullet asleep on the sand wait for the solider pelagic species such as pampano to come in. At the extreme edge of the flashlight's beam a pale shape is glimpsed for a second. It might have been imagination or else a momentary fault in one's retinal wiring (pressure does strange things to night vision). It is worth pursuing, though, and one holds the beam on the spot where the object may have been and makes a burst of speed with both plywood flippers. At night most fish are either immobilised by darkness or else vanish with a fin's flick; pampano are strange in that they seem to allow themselves to be pursued, partly alarmed and partly attracted by the light. They could easily escape, but often after an exhausting chase one can overhaul them. In the light they are round and silver, about the size of a dinner plate and, in contrast to most coral species there is good meat on them. The technique is to hold the torch at arm's length and out to one side. Like all laterally flattened fish, pampano turn so as to present an attacker with an edge-on view, but this attacker has out-thought it and for a moment it is almost sideways on. If the aim is true and the speargun works that moment should be enough. The range is never more than seven or eight feet. More than that and the spear will not penetrate. Twice that distance is the effective limit of visibility with a two-battery torch. A brace of pampano (for if there is one, there will be others) is plenty. On a good night one may be out and back and building a fire within forty minutes.

In much of South-East Asia the pressures of virtually unregulated commercial fishing have led to close parallels between the fishermen of villages like 'Sabay' and those of Fraserburgh. In Scotland it is—incredibly—not illegal to trawl right up to the beach. This practice has done enormous harm to littoral fish stocks, as to the locals who in calm weather could once go out in small boats and fish safely within a few hundred yards of shore. In the Philippines the equivalent is provided by the proliferation of *buli-buli* and *basnig* fishing. *Buli-buli* refers to large seine nets of very fine monofilament mesh, often as small as ten millimetres. *Basnig* uses similar nets but at night, with banks of bright lights to attract squid and nocturnal species. The older craft use propane gas for their lamps, the modern ones electricity. A *basnig* fleet with its cityscape of lights and distant massed chugging of generator engines is a characteristic sight. From its deployment one can often tell as much about local politics as about the offshore reef formations keeping it at a distance. Officially, there is a seven-kilometre limit inside which only 'sustenance fishermen' may legally fish. In addition, Fisheries Administrative Order no. 164 places restrictions on all *buli-buli* fishing. It is illegal to use a boat of more than three gross tons, as it is to use a net whose stretched mesh is less than 29.9 millimetres. Smaller boats using legal nets may fish within the severn-kilometre limit but come into the jurisdiction of the local municipality. Violent battles sometimes erupt between *basnig* and *buli-buli* fishermen on the one hand and locals on the other who claim their livelihoods are being ruined. They are not wrong; inshore fish stocks are visibly depleted after a single night's close approach by illegal fishermen.

As in Scotland, there is the same acknowledgement that unpoliced practices combined with over-fishing can drive small fishermen to crime in order to stay alive. In many provinces there is a steady battle between thinly stretched, and sometimes conniving, local authorities and an army of people who go fishing with home-made dynamite, cyanide and bleach. In fresh water, electricity is also used. One can occasionally see small men staggering about in a shallow river with a 12-volt Jeep battery strapped to their backs, prodding the water with two terminals. Poison is mostly used to stun reef species for export as aquarium fish. Those it does not kill outright it weakens, and it is estimated that maybe three-quarters die within a fortnight. It also kills coral polyps. So does dynamite; and the skill with which it is often both made and used does not alter the fact that non-target species die as well, and in deeper water retrieval rates may be fairly low.[1] But as local fishermen—who are neither blind nor stupid—will say, what can you do when you have a family to feed and fish are so scarce?

In all this anarchy there is one thing to be said about fishing at local level in South-East Asia: precious little goes to waste. There is hardly a species which is not eaten, nearly nothing too small to eat. In the case of fried fish, much of the skeleton is often eaten as well. Even the tiny conical hearts of certain mackerel-like *Scombridae* are saved. To this extent local fishing in the developing world is free of the cavalier squandering and disdain which accompanies commercial fishing by the wealthy nations. To watch the fishermen of 'Sabay' and their families handling fish, whether alive or dead, is to witness a radical respect for food.

At the extreme opposite is the deep-frozen brick of supermarket cod, prawns, sole fillets, tuna steaks. This sanitised object represents merely the pinnacle of an industrial pyramid of slaughter, destruction and waste. To speak with refined, Western sensibilities in mind: in terms of seemliness it is no longer possible to propose fish-eating as somehow less objectionable than meat-eating. In terms of ecological damage, the worldwide plundering of marine life may turn out to have been even more disastrous than the felling of rain forest for the benefit of beef ranchers.

---

[1] For a fuller description of these methods see James Hamilton-Paterson, *Playing with Water* (1987).

II

'The beauty been'

Popular conservation has helped promote the notion that there must have been an ideal 'balance of Nature' which existed before *Homo* began upsetting it. This is nonsense, of course. The history of life on Earth is full of episodes like the great Permian extinctions at the end of the Palaeozoic. These not only wiped out the ubiquitous trilobite, which had survived since the Cambrian, but about ninety-six per cent of all living species of fauna as well.

If there is an unfortunate consequence of James Lovelock's elegant and serious 'Gaia' thesis, it is that it lends itself to being hijacked by all manner of fringe theorists and used to support their own dotty notions. Thus, the idea that the biosphere might in some sense be self-regulating so as to maintain conditions favourable to life has been peverted, turning the planet into a sentient Mother Earth. Anthropomorphising it as 'she' makes 'her' struggles to maintain an immemorial balance in the face of *Homo*'s despoliation seem intentionally remedial, even noble. In this reading, 'Gaia' comes across as a somewhat sainted landlady, trying her utmost to accommodate her latest lodgers who have turned out to be slobs and vandals intent on ruining her delicious mansion. Such an idea is absurd and unscientific and should not be blamed on James Lovelock. It is what happens when people full of millennial *Angst*, guilt feelings and moralised views of biology adapt a serious hypothesis for their own purposes. They are dealing with a goddess, or maybe an old tyrant who once let all but four per cent of her animal lodgers die out. We are dealing with evolution. Conditions on this planet are constantly changing, and always have been. The extinction and evolution of species is as mutually dependent and constant a flux as the Earth's crust is itself plastic, wobbling to the belches of vulcanism and to the tuggings of celestial bodies. *Homo sapiens sapiens* is only one of an estimated 30 million species, and it is possible to argue that since he is as 'natural' a creature as, say, the crown-of-thorns starfish, *Acanthaster planci*, any results of his presence, no matter how devastating, are also 'natural'.

There is often a blurring of motivation in public 'Green' campaigning which amounts to a principled dishonesty. The millennial perception is that *Homo*'s remodelling of the balance of the biosphere has fatally endangered everything. The astutely televisual selection of species of cetaceans and large mammals (whales, dolphins, elephants, pandas) has actually come to stand for the sudden awareness of *Homo*'s threat to *himself*. In recent years there has been a panicky rise in the number of TV and radio programmes, magazine articles and scientific papers devoted to the physiological and psychological threats to man's continuing existence. In the first case it is constantly repeated how menaced he is by his own poisons, effluents and by-products, as well as by his thoughtless exploitation of delicate global structures. In the second, it is a darkly lurking shibboleth that urbanisation leads inexorably to a breakdown of social behaviour, to mental illness, endemic envy and dissatisfaction, to out-and-out psychosis and mass murder ('Just look at what's happening in America . . .'). Man is perceived as being threatened simultaneously from without and within. The timely adoption of lovable big animal species to gather like a lens the varied gazes of his self-concern cannot be dissociated from the approaching millennium. We reach for pandas as a child for his teddy.

It is entirely proper to wish to preserve every species on

Earth, whether (to our eyes) magnificent creatures or humble slime-moulds, and not only because of their interdependence. *Homo* is an aesthetic animal and has pungent notions of the sublime. He feels diminution when the familiar vanishes. He experiences with the greatest upset the ghost of loss which stalks his waking days as well as his sleeping hours. It mobilises in him a tenderness akin to vulnerability, to the point where a large part of his wistful concern for whales and the environment generally is displaced fear for himself, an intense longing looking for somewhere to alight. It becomes vital for him to know that whales and wilderness still exist somewhere on Earth, even if he never sees or experiences them at first hand. They represent cardinal points on the map he has inherited from his ancestors, representations of an earlier world where such things fixed the terms of daily living. Obscurely he feels that without them he is lost, or at least that he cannot shed them without consequences he can neither foresee nor articulate. If only he would say so! The high principles are there, to a fault; but so is dishonesty, because in shifting attention on to certain choice species he postpones recognising his poignant lack, his own strange delicacy. To this extent *Homo* needs more self-interest and not less, provided it attends to that interior chart, to the space which is the ancient and common legacy of his species and whose territory he necessarily inhabits.

What alarms him most and makes him most confused and despondent is the speed at which environmental changes have taken place. This is a comparatively new phenomenon. It is not until the late nineteenth century that a sustained note of mourning is heard for parts of the familiar scenery that have vanished. Since it had to do with a rapid rise in industrialisation and population this note was first heard in Britain. It was typically sounded by Gerard Manley Hopkins in his 1879 poem 'Binsey Poplars,' on the cutting down of a row of trees near Oxford to make way for housing ('After-comers cannot guess the beauty been'). Thereafter it is heard increasingly, quite distinct in tone from a generalised literary *eheu fugaces*. The snows of yesteryear are, after all, nearly always followed by the snows of this year and those of next. Hopkins was recognising something new: the destruction of trees that never would be replanted.

Since then, this process has accelerated so that today only the most purblind twenty-year-old could fail to notice a vanishing of things even since his own childhood. Those twice his age and more may have the vividest childhood memories of entire landscapes which are now gone, while people brought up in horsedrawn days before World War I can find themselves in alien terrain. This is a disturbing thing to have happened and leads to the ironic conclusion that it may often be easier to deal psychically with sudden disaster than with steady attrition. French and Belgian farmers whose fields were churned into wastelands by trench warfare knew they had only to wait a few years before most of the greenery sprouted back again. Conversely, those people who grew up to the rich, tangled hedgerows and small fields of the British countryside in the 1930s, 1940s and even 1950s know they will never see them again, nor many of the flora and fauna associated with them. The process is mostly irreversible, not because the species would not return if allowed to or—like a row of poplars—were replanted, but because roads and housing are never ploughed up but only proliferate.

Looking back, it strikes us as curious how seemingly unaffected people once were by the extinction of a species within their lifetime, as though they were isolated zoological events without particular consequence and foreboding nothing. Only now and then is it possible to detect misgivings beneath the breezy pragmatism, even sometimes a current of unease. 'Last chance to see what Audubon saw,' wrote an unnamed columnist standing in front of a cage in Cincinnati in 1913. This contained 'Martha,' the sole remaining passenger pigeon in the world. This species, *Ectopistes migratorius*, had within living memory been the most familiar bird in the whole of North America and was prettily illustrated by Audubon in his *Birds of America*. Its wholesale slaughter by settlers peaked in the mid-nineteenth century with massacres involving entire flocks of tens of thousands of birds. The last flock of this handsome creature—it was somewhat larger than an English woodpigeon and beautifully marked—was sighted in Illinois in 1895. The last recorded wild specimen was shot in Quebec in 1907. 'Martha,' meanwhile, had been bred in Cincinnati Zoo in 1885. She died at 1 PM on 1 September 1914, aged twenty-nine, and with her an entire species. 'The main cause of their extinction is plain for all to see,' wrote Richard Carrington some forty years afterwards. 'Men cannot escape the moral responsibility for the callousness, the greed and the supreme irreverence for life that led to the passing of the passenger pigeon.'[1]

This particular bird had followed many other species, which included the dodo—hunted to extinction in Mauritius in the late seventeenth century—and the great auk. This last, the garefowl, was also flightless. It was hunted in the North Atlantic for its fat, sometimes being killed because it was associated with superstitious belief. One was burnt in Ireland in 1834 for being a witch. The very last specimen was killed ten years later. Today in New Zealand heroic efforts have been made to save the kakapo. This large green-and-yellow bird, something between an owl and a parrot, cannot fly, never attacks and never defends itself. When it can be encountered at all (for, scarcity aside, it relies on exquisite camouflage) the kakapo has a further claim to be remembered as an example of Nature's diversity for it smells strongly of freesias. Aesthetics apart, the loss of a species has consequences which can never

---

[1] Richard Carrington, *Mermaids and Mastodons* (1957).

be entirely predicted. The extinction of the dodo eventually had effects on the very landscape of Mauritius, certain hardwood trees failing to seed themselves and beginning to die out. Only in recent years has it been established that the dodo played a central role in the seeding and distribution of these trees by eating their fruit. The bird's digestive tract softened the seeds' testa and assisted germination. Now turkeys are being imported from America in the hopes that they may act as substitute dodos.

While endangered species embody poignant reminders of our own mortality, it is the vanishing of entire landscapes that upsets us most. There is nowhere left to turn for solace and with which to recreate the continuity of our lives. Sights, smells and sounds may all vanish. I pretend not to mourn the wild profusion of the natty yellow-and-black striped caterpillars of the Cinnabar moth which once stripped to the bone the clumps of groundsel to be found on every patch of wasteland in southern Britain. Likewise, I miss the sheer variety of other butterflies and moths (including many rare species) which appeared even in the most suburban garden as late as the early 1960s. One knew where Yellow and Red Underwings would be and when in May to look for the Angle Shades moth just after it had hatched and its colours were at their freshest. The subtlest peach and brown and olive tints seemed to hover a fraction above the surface of its wings as if pure colour stood off its scales by the thickness of dust, glowing and velvety. It now seems both important and hopeless to wish for other people such pleasure and ravishment, whether of looking at moths or being frisked around by dolphins. It is true that after-comers can never know exactly what they have missed; but missing things in our own lifetimes sets in motion the inarticulate hollowings of loss, and in turn we apprehend how quickly ordinary beauty is being made to vanish as if the hand of man held a wand whose touch made some things disappear for good and turned all the rest to lead. Each generation adapts to an impoverished world, but for the first time people are conscious of having to make do with remains. This has its effects.

In 1918 a steamer was wrecked near Lord Howe Island, about four hundred miles off the east coast of Australia. Its rats swam ashore and changed the landscape for ever. Two years afterwards one of the islanders wrote, 'This paradise of birds has become a wilderness, and the quietness of death reigns where all was melody.'[1] Such events have become common on islands, where casually or accidentally introduced domestic animals (goats, rabbits, cats and dogs) as well as vermin can decimate native species and upset an entire ecology.

Of all the laments for a vanished landscape, perhaps that by Edmund Gosse tells Britons most vividly what they have done and what they may no longer see. No chance shipwreck caused the disaster. On the contrary, if anything it was caused by zoology allied to a strong religious bent. At the ages of nine and ten Gosse spent countless hours with his famous naturalist father examining rock pools on the Devonshire coast, 'a middleaged man and a funny little boy.' His account was published in 1907 and refers to the years 1858 and 1859, immediately after the publication of *Omphalos*. In the intervening half-century the whole seascape changed. The passage is a beautiful description of a lost world, and one of the first examples of a sensitive and passible man fully aware of what has gone and how its going was part of a pattern destined to be repeated at ever-gathering speed. In its prescience, it is a warning of the possible consequences of scientific fieldwork, even of 'eco-tourism'.

If anyone goes down to those shores now, if man or boy seeks to follow in our traces, let him realise at once, before he takes the trouble to roll up his sleeves, that his zeal will end in labour lost.

[1] Quoted in Rachel Carson, *The Sea around Us* (1951).

There is nothing, now, where in our days there was so much. Then the rocks between tide and tide were submarine gardens of a beauty that seemed often to be fabulous, and was positively delusive, since, if we delicately lifted the weed-curtains of a windless pool, though we might for a moment see its sides and floor paven with living blossoms, ivory-white, rosy-red, orange and amethyst, yet all that panoply would melt away, furled into the hollow rock, if we so much as dropped a pebble in to disturb the magic dream.

Half a century ago, in many parts of the coast of Devonshire and Cornwall, where the limestone at the water's edge is wrought into crevices and hollows, the tide-line was, like Keats' Grecian vase, 'a still unravished bride of quietness.' These cups and basins were always full, whether the tide was high or low, and the only way in which they were affected was that twice in the twenty-four hours they were replenished by cold streams from the great sea, and then twice were left brimming to be vivified by the temperate movement of the upper air. They were living flower-beds, so exquisite in their perfection, that my Father, in spite of his scientific requirements, used not seldom to pause before he began to rifle them, ejaculating that it was indeed a pity to disturb such congregated beauty. The antiquity of these rock-pools, and the infinite succession of the soft and radiant forms, sea-anemones, sea-weeds, shells, fishes, which had inhabited them, undisturbed since the creation of the world, used to occupy my Father's fancy. We burst in, he used to say, where no hand had ever thought of intruding before; and if the Garden of Eden had been situate in Devonshire, Adam and Eve, stepping lightly down to bathe in the rainbow-coloured spray, would have seen the identical sights that we now saw, – the great prawns gliding like transparent launches, anthea waving in the twilight its thick white waxen tentacles, and the fronds of the dulse faintly streaming on the water, like huge red banners in some reverted atmosphere.

All this is long over, and done with. The ring of living beauty drawn about our shores was a very thin and fragile one. It had existed all those centuries solely in consequence of the indifference, the blissful ignorance of man. These rock-basins, fringed by corallines, filled with still water almost as pellucid as the upper air itself, thronged with beautiful sensitive forms of life, – they exist no longer, they are all profaned, and emptied, and vulgarized. An army of 'collectors' has passed over them, and ravaged every corner of them. The fairy paradise has been violated, the exquisite product of centuries of natural selection has been crushed under the rough paw of well-meaning, idle-minded curiosity. That my Father, himself so reverent, so conservative, had by the popularity of his books acquired the direct responsibility for a calamity that he had never anticipated became clear enough to himself before many years had passed, and cost him great chagrin. No one will see again on the shore of England what I saw in my early childhood, the submarine vision of dark rocks, speckled and starred with an infinite variety of colour, and streamed over by silken flags of royal crimson and purple.[1]

In its way this is a classic threnody, indulgently melancholic even to the furtive pleasure of 'No one will see . . . what I saw.' It also sets out the biblical coordinates of a mode which survives to this day as a shadow between the earnestly secular lines of much conservationist rhetoric: 'the Garden of Eden,' 'paradise,' 'vision,' 'profaned' and, of course, the Father. This is touching in its confusion between the child who could not always distinguish his father from God, and the adult writing a devastating biography who could see the difference all too clearly. Both personae are there in this passage, unconsciously revealed by Gosse the son in two scientifically conflicting phrases: 'undisturbed since the creation of the world' and 'cen-

[1] Edmund Gosse, *Father and Son* (London, 1907).

turies of natural selection.' The first represents his father's firm belief (he was a devout creationist) while the second is entirely his own, convinced Darwinian that he is. This identification of his father with the Creator, and the *ur*-landscape of his childhood in Devonshire with Eden, maybe gives an additional clue as to why we can feel so devastated by the disappearance of the place where we grew up. Landscape blurs easily into the parental.

This mourning for landscape, this apprehension of death without a proper body to grieve over, is one of our modern era's cruellest legacies. It has often been recorded but seems to have gone largely unremarked as a likely cause of common forms of despair and depression. For lack of any medically plausible origin these are presumably attributed to the usual domestic disorders, disappointments and jiltings; whereas in reality the sorrow may be far grander, more pervasive and unsolaced, its cause misunderstood by both sufferer and doctor.

It is said that the British reading public's nostalgia for the imagined certitudes of Victorian and Edwardian England accounts for the popularity of rural diaries and reminiscences. This may be so; but it is quite as likely that the nostalgia is as much for a lost landscape as for any vanished social order.[1]

---

[1] Sooner or later this ailment will doubtless be officially recognised and accorded its own name—probably an inflated tag along the lines of 'Topological Mourning Syndrome' or even 'TMS.' Actually, the condition itself is by no means always vague, and can present quite specific symptoms. 'It makes me feel older,' is how a middle-aged villager in 'Sabay' described viewing the Fantasy Elephant Club across an otherwise familiar strait.

III

MARGINALIA:
Parrotfish

Frozen parrotfish imported from the Caribbean are for sale in south London fishmongers. The *Scaridae* are specifically reef fish: the beak-like fusing of the front teeth which gives them their popular name has evolved so that they can browse on corals. The parrotfish does indeed damage corals, just as a blackbird damages worms. What counts is the balance of such activities, especially in an ecosystem as delicate as that of a tropical reef. Pulling parrotfish out of reefs in quantities sufficient to supply an export industry cannot but have an unbalancing effect. Through a complex chain of relationships, the reef will inevitably suffer from their absence, as it certainly will from the methods used to catch them.

This is doubly exceptionable since the members of this family make indifferent eating except at the moment they are caught. Then the otherwise uninteresting meat is delicious raw, marinaded for fifteen minutes in lime juice, oil, tiny native onions and black pepper. Frozen into woody curves in a bin in a Peckham fishmonger's, a parrotfish merely pretends to the exotic, its blunted colours ghosts of what they were. Who knows whether their corpses are there to satisfy the nostalgia of an immigrant population or the local consumer's fickle desire for novelty. Either way, a reef off Barbados or Jamaica is now undergoing a change which will most likely be permanent.

The ocean's emptiness appals the swimmer, but only because it can supply nothing for his own survival. He cannot entertain flabby polemic about dolphins. His is the mind of a man lost in the sea. Yet even as he struggles to save himself he is hollowed out by despair. What is it that he is saving? The thought corrodes his every intention. In this wide salt world which he treads he is nothing, has nothing but a face mask and a pair of trunks. Until one loses everything it is never clear what it was one had. Now, in a bleak inner glimpse, he finds he has dissolved. The landscape of his own past, his private history, seems to have vanished, leaving only a sense of attrition. As he glances down through the water his body dwindles whitely like a distant peg and sheds a small discoloured puff of urine which briefly unravels itself in thready convections like those of lime juice being diluted. Nothing but ocean. His entire body is dissolving, too. He only ever existed as three-tenths and that fraction is melting into water.

However, this thirty per cent contains an animal which does not want to die. A passive animal, maybe, but still perversely convinced that help will turn up as if by more than mere chance. Sooner or later someone surely has to pass within hailing distance of the psychic beacon he must have become, broadcasting his distress signal on all frequencies. He squints at the sun. Now that he no longer wants it to be stuck vertically at noon it seems reluctant to move at all. Night with its hope of fishermen is still many hours away.

The swimmer tells himself he need not bank only on them. He has been overlooking all the other sorts of boat which continually cross these waters. Besides tattered inter-island launches, there are all the craft which used to fetch up on 'Tiwarik': friendly gunrunners, wanderers from the south with their faces wrapped against the sun, poverty-stricken vagabonds neither peaceable nor violent but chance-takers of more or less competence. Any of them might spot him from miles away with a vulture's quick eye for a weakening beast. He tries to imagine into being a huge arch of cloud letters in the sky: reward!, and underneath a gigantic arrow pointing straight down whose tip balances on his sunburned head. It is a message aimed impartially at any of the seagoing mavericks who still inhabit this last corner of the ocean.

So hard does he will it that he soon thinks he hears, above the infuriatingly loud slop of wavelets, the faintest putter of an engine.

7. Pirates and Nomads

Axel Heyst's 'magic circle' would just have enclosed Manila, Saigon, Singapore, Surabaya and Ternate, as well as the Sulu and Riaw-Lingga Archipelagos. (The Mergui Archipelago lies somewhat outside.)

I

It was a morning of flat calm in the Sulu archipelago towards
the end of November 1990. The sun in its early angle was
both gentle and powerful, forcing a luminous violet into the
water. The islands dotting the sea to the horizon stood clear and
delicate, as though the same gentle power had hatched them out
overnight and left them in their freshness to harden off.

A small boat with bamboo outriggers was heading back
towards Subuan. In it were three young Bajau women and a
double row of grey plastic jerrycans of water. Perched on these
amidships was a seventeen-year-old boy, the brother of one of
the girls, who understood the ten horsepower Briggs & Stratton
engine. He was controlling the speed by means of a length of
nylon fishing line tied to the throttle. This ran over a rubber disc
cut from the sole of a sandal and nailed to a thwart. The boy had
wrapped the end of the line around his finger and was concentrating on holding the engine's note at the same pitch. It was a
nearly new motor whose exhaust was rigged in local fashion
from an iron elbow bolted to the exhaust port with a foot of
ordinary water pipe screwed into it and jutting up at an angle.
From this unsilenced tube blared the fumes and racket familiar
to those on board as a soothing part of a ritual which, before the
coming of the engine, had been a much more laborious business
involving paddles and a sail made from rice sacks.

The Bajau were homeward bound from the island of Siasi,
whose town was the nearest reliable source of fresh water.
There was also an acute shortage there, to be sure, but at least

water could be bought at a seasonal average of about twenty cents per jerrycan. At first light they had reached Siasi's jetty with a load of fish, which the women's husbands had caught the previous day and had kept alive overnight in rattan containers suspended in the tide beneath the floors of their huts. All the Bajaus' transactions had been carried out by the one woman among them who knew how to use money. None of the four was literate, but she did understand coins. As long as everything was done in coins it was all right. She would have nothing to do with paper money. Not only could she never be sure of the denominations but it was all too evidently flimsy stuff. Drop it in the sea and it would be reduced to pulp and where were you then? For all her knowledge, though, she and her companions were happiest with barter. Barter, now, that was the proper way of doing things; taking into account all sorts of subtle variables like quality, whether there was a glut or a shortage that day, how much of a hurry you were in. Thus a basket of medium blue crabs might fetch two baskets of good cassava, while a 'half-boat' shark half the length of the *bangka* was worth quite an assortment of nylon line, fish hooks and petrol, plus (especially if the Chinese shopkeeper was buying jaws as well as fins) two pairs of children's shorts.

As the boat rounded an islet they could see Subuan in the distance with its detached clump of huts standing well offshore on legs anchored in the corals. The sea's surface glassily collected the white clouds which, towards midday, might with any luck heap together and wring some precious rainwater out of themselves. The Bajau stared forward, reading the water. Probably no other people anywhere could gaze with such knowledge of what was happening below its surface. The sudden swirl of a tail, catspaws and dimples of wind, the alignment of the snapped blades of *Thalassia* sea grass whose floating debris threaded the archipelago like oil slicks, even passing smells: all carried information or stood for omens.

No doubt the hammering exhaust and the mesmeric calm prevented the women and the boy from noticing the bigger and faster craft with double outriggers slipping out from behind the islet as they passed. It always would have been too late for them in any case. They would certainly have seen it as it drew level, matching their speed a few yards away. The boy at the engine would have turned, half stood, shocked but not really bewildered by the sight of a man steadying an M-16 across the bigger craft's low roof. And thus, probably without shaping a clear thought, he took leave for ever of his sister and companions, of the glittering morning ocean, of his seventeen years. The shots carried away much of his head. His body must have fallen across the scorching engine, for when it washed ashore it was still possible to make out extensive blistering of chest and stomach. His dead finger slack, the engine would have slowed abruptly to tick-over. The little boat lost way at once, rocking gently out in the middle of the sunlit strait, steam rising as the boy's blood hissed and bubbled on the cylinder head.

It is unlikely the three women remained sitting mutely while men swarmed aboard across the outriggers. Being Bajau, their instincts would surely have been to put themselves into the hands of the sea and its spirits. But sooner or later, no matter how ably they swam and dived, they would have been wrenched from the water and tied up on the larger craft. A man aboard their *bangka* unwound the nylon from the boy's finger, hauled his sizzling body off the engine and threw it overboard. Then the faster boat took the other in tow and, veering away from Subuan, probably headed south-west to Tapaano or even Sugbai, which lay on the horizon a scant forty minutes away.

Once there, the pirates would doubtless have been joined in camp by their outlaw companions, taking it in turns to rape and beat the women unconscious. At length someone would have set off with all three victims' sarongs (that all-purpose garment which serves variously as skirt, trousers and turban) and arranged to have them delivered to the respective husbands, together with ransom demands of just under $1,500 per head.

Since everyone knew the Bajau were nearly all subsistence fishermen and the poorest tribe between Mindanao and Indonesia, it is not clear how seriously these ransom demands were intended. Maybe the pirates thought that if a poor Bajau could put an engine in his *bangka* he had a secret source of funds which might be further tapped. In point of fact the motor had been installed by a Chinese, a merchant keen to boost his trade in marine products which he periodically shipped to Manila. In any event he was not a man to throw good cash after a lost engine. After all, no one seriously thought a Bajau was worth . . . well, how could one put a figure on the lives of illiterate fishermen, gypsies who mostly lived outside any sensible economy? Still less on their wives.

And so, after a week's captivity, during which time no money arrived and their treatment no doubt reflected the pirates' increasing anger, the three Bajau women were put to death in a way which can only be imagined. All that is known is that the body of one was chopped up into quite small pieces—*diced*, it would be called in a recipe—and piled in the bilges of their little *bangka* over the four bolts on which the Briggs & Stratton engine had been mounted. Then it was set adrift. The boat with its heap of meat wandered with the current for a while before being found, recognised and read as just one more awful warning in a region used to awful warnings and worse deeds.

This small atrocity was still mentioned now and then by the island folk three months later, although by that time it had been overtaken by news of more recent acts of piracy. Such things were commonplace. The least advantage in material goods or business put anyone firmly in an extortioner's sights. Uncharacteristically, these pirates had forgone a *bangka*, but they had gained a nearly new engine, some full containers of drinking water and a week's fun. The world spun on.

As for the Bajau themselves, they would not have forgotten. At least they were able to take the boy's burned and headless body to their cemetery island and lay it with semi-pagan rites beneath a carefully painted wooden board, surrounded by gay little flags. Then, as is their custom, they probably raced each other back down the beach and into the water to wash, spurred on by the half-serious belief that 'the last one in is the next to die.' Perhaps they instinctively felt the sea safer than the land. Maybe some distant tribal memory warned them that, like turtles, they were doomed to transact certain necessary rituals on dry land before they could once again return to the comparative safety of the ocean.

Such an anonymous event—which was never reported in any newspaper, nor formally to any military or civil authority— contained within it many of the well-worn coordinates of the Bajaus' fate, not merely of three women and a teenage boy but of the whole scattered tribe. Come to that, it was characteristic of an entire region. Everything that had taken place was immemorial within its setting: the fetching of water, the bartering of goods, the being victim, the being pirate. So also were the lumps of land rising from the seabed haphazard of all demarcations, the shoals and atolls, sandbars and islets drugged with sun as the archipelago (a word whose beautiful syllables stretch themselves to the mind's horizon) sprawled in its great tropic swoon while seething with violence. Many of Conrad's best stories were set in this area. From southern Mindanao, from Sultan Kudarat and Zamboanga and Palawan down to Java in the south, from Sumatra in the west to the Moluccas in the east, certain things have changed little to this day. Some names are different. Celebes has become Sulawesi, Makassar is Ujung Pandang and Batavia is Jakarta. But the rest—Ternate, Surabaya, Kuching, Samarang, Timor—still exist and echo with the pungency that thrilled my adolescence until like Axel Heyst himself I could believe my life enchanted by a magic circle 'with a radius of eight hundred miles drawn round a point in North Borneo.'[1]

The way of living is still much the same for many in the Sulu archipelago. They have not stopped diving to eardrum-splitting depths for pearls and *tripang* (sea cucumbers); the blue highways of water are patrolled as ever by sharks and crisscrossed by inter-island traffic of every kind. The pirates, smugglers and common cut-throats are very much in evidence. Even vestiges of the old sultanates remain, although the turbaned despot living in his fiefdom—a pocket trading empire defended by mangroves, a treacherous estuary and riverbank spies—tends nowadays to be a pretender who manages a grocery store in Jolo and writes long letters by candlelight to the United Nations, beseeching it to recognise him as rightful heir. Everywhere the *kris* has given way to the Armalite, while the great white sails of clippers and schooners have been supplanted by Isuzu marine diesels. Otherwise, Conrad might recognise these waters as having retained their archipelagic essence: seductive, dangerous, possessing above and below their surface a treacherous quality which leaves nobody untouched.

He would certainly have been familiar with the Bajau, although in his day they were more exclusively boat-dwelling than they are now. Anthropology has still not decided where they originally came from, or why. The earliest European visitors, beginning with the Portuguese and Magellan in the first quarter of the sixteenth century, reported small groups of

[1] Joseph Conrad, *Victory* (1915).

boat-dwelling nomads throughout the archipelagos of South-East Asia. It is largely a matter of guesswork as to how long they had been there. Opinion is divided, too, on whether they once all came from the same tribal stock—whether, for example, the sea nomads of the Mergui archipelago off the western coast of Burma share a common ancestry with those of the Riaw-Lingga archipelago in Indonesia and the Bajau of Sulu and North Borneo. Whatever their origin, it is always presumed that at some point far back in time they were all land-dwellers who for some reason decided to go and live, as far as possible, afloat. What is now to be seen is the final phase of this ancient way of life as the Bajau try to take up residence on land again. The difficulty they are experiencing in doing so is a measure of the thoroughness of their social adaptation to living on the sea.

One of the theories as to why their ancestors left the land in the first place is that they were literally driven off it by stronger tribes. If so, it is ironic that similar persecution is now driving the Bajau off the sea. Three tribes predominate in Sulu: the Tausug, the Samal and the Bajau. For several reasons, principally linguistic, modern opinion tends to lump these last two together. The Bajau language is a dialect of Samal (properly called Sinama) and besides, they consider themselves to be Samals of a kind. 'Of a kind' is a reasonable qualification since the Samal of the southern archipelago are a heterogeneous bunch and can vary in dialect, attitudes and social customs from island to island. At one extreme they include illiterate, pagan, boat-dwelling Bajau and at the other sophisticated traders and teachers who have made the *hajj* to Mecca.

Ranged against all of them are the Tausug. The Tausug have always been politically and otherwise dominant in the archipelago, with a reputation for pride, hot temper and general disdain for lesser folk such as Samal. As for the Bajau, they often refer to them as *luwaan. Luwa* is the Tausug verb for 'to spit out'; 'outcast' would probably be the nearest English equivalent. No doubt a good deal of Tausug bigotry towards the Bajau is caused by—or at least explained as—a matter of religion. The Tausug have a generally high opinion of their own version of Islam, rather less of the Samal version, and can express serious doubts about whether the Bajau brand even counts as Islam at all. It would be impossible for a people living in a predominantly Islamic area not to have assimilated a great deal of Islamic culture, and the Bajau have done so, many of them being devout Muslims. But since they are traditionally a boat people they are perforce a people without architecture. Hence their mosques—for all the world dilapidated huts on stilts standing in the waves—do not fulfil the Tausug idea of what a mosque should look like. 'How can a people without decent mosques qualify as Muslims?' I was asked by some exceedingly hospitable Tausug on Siasi. What I should have countered with, had I not lacked the requisite nerve and discourtesy, was 'How can a people drink as much as you do and qualify as Muslims?,' for drunkenness in this famously Muslim region is widely admitted even by imams to be one of the principal reasons for the extraordinary level of casual violence. It was, of course, Tausug pirates who killed the three Bajau women and the boy. Although it is quite certain their Bajau relatives could identify the men involved, they would have been much too frightened to report them even had there been any form of law in the area worth reporting them to. Seen from far enough away, of course, the Tausug and the Samal and the Bajau appear to have far more in common than not—seem practically indistinguishable, in fact, since they are all people whose lives are overwhelmingly mediated by the sea.

Left to himself, the Bajau's is a peaceful, shy, somewhat nomadic way of life which is as highly specialised as that of the

Eskimos. It would be a mistake to romanticise it, though. Bajau living has always been extremely hard, often dangerous, lonely and beset with disease. The traditional family unit was a married couple and their children in a low-roofed boat. Depending on season, preferred fishing grounds, family events (mainly marriages and deaths) and sheer whim, the boats either formed parts of flotillas, floating villages or went their own separate ways for months on end. They tended to put ashore only for drinking water, firewood and cassava; otherwise it was a life at sea in all weathers. It was also a life largely spent crouching: squatting to cook, squatting to eat, squatting to fish. To this day older Bajau can still be seen who walk on land with a gait as characteristic and graceless as a duck's, their lower limbs slightly atrophied from a lifetime's hunkering down.

The conditions in these boats, where they still exist today, can become squalid, to say the least. It is not merely that if there are infants and toddlers aboard the bilges soon give the craft the atmosphere of a floating urinal. All sorts of vegetable ends, cooking scraps and fish guts fester there as well, brewing up in the tropical climate so that the liquid, when glimpsed between plank and thwart, can be seen fizzing. This is baled out at anchor in shallows, but on the high seas is often allowed to build up, and for a perfectly good reason. Since Bajau men spend a good deal of time overboard cutting seaweed and diving for *tripang* and pearl oysters the last thing they want to do is lay a trail of offal and attract sharks. This would be a determining factor for any sea dweller in the tropics and, indeed, is noted in a curious account by an Englishman, Leopold Ainsworth, of trying to run a timber business in the Mergui archipelago in the 1920s.[1] The sea gypsies he knew there, and

---

[1] Leopold Ainsworth, *A Merchant Venturer among the Sea Gypsies* (London, 1930).

whom he tried to introduce to the idea of paid labour on land, called themselves 'Mawken,' which he translates as 'sead-rowned.' The boats and habits of the Seadrowned People of the Mergui archipelago seem to have been not much different from those of the Bajau, which is hardly surprising given the similarity of the conditions. The Mawken, too, gave an account of their origins which told of persecution by Burmese hill tribes from the north and Malay pirates from the south squeez-ing them off the land and into the sea. They also offered this as an explanation of why they had no interest or skill in cultiva-tion, merely picking up fruit or an occasional wild pig when passing an island. According to Ainsworth the Mawken seemed to exist on seafood, rice and opium. The Sulu nomads mostly eat cassava as their staple, only the wealthier tribes and classes relying on rice. I never saw opium, but marijuana is widely smoked.

Malnutrition is a common consequence of this way of life, partly because the diet is unvarying but also because the choic-est fish are mostly reserved for sale while the fisherman himself subsists on scraps, shells and the coarser varieties. Combined with the lack of sanitation this leads to a high incidence among the Bajau of tuberculosis, malaria, infant diarrhoea and infes-tations, to all of which their resistance is slight. This seems ironic, even contradictory, in view of the extraordinary physi-cal fitness and imperviousness to cold needed by men free-div-ing to up to one hundred metres for pearl oysters. Even greater depths are claimed, and although so far as I know nobody has ever bothered to measure them exactly, such dives have to be placed against the current world record of 105 metres,[1] espe-cially since these are working dives during which shells, lob-sters or sea cucumbers are gathered. Before diving the men often swig a mixture of canned milk beaten up with little

[1] See p. 115.

'native' eggs and lemon juice and eat bananas. They say this makes them resistant to the cold.[1]

Young Bajau, particularly children and adolescents, often have the bleached, tawny hair of people who spend their lives in and out of the sea. Some even approach a dusky blondness which, amid uniformly black Asian hair, is very striking indeed and a sure guide to that person's way of life and social status. No doubt it is Western culture, and specifically cinematic images of Californian or Australian beach culture, which equates this with the very pinnacle of healthy living (or did before skin cancer was mentioned). In South-East Asia, though, it speaks of a life of poverty and often malnutrition lived beyond the least shadow of a classroom, with intestinal worms and scabies as constant companions.

Even in the 1960s, when anthropologists like H. Arlo Nimmo were undertaking classic studies of the Bajau, their landward trend was obvious. The entirely floating life on houseboats was being relinquished for a tentative one in stilt huts which represented the placing of a wary toe on the very edge of land. Like their Samal cousins, certain Bajau always did build communities of huts on stilts in shallow offshore areas, each hut joined to the next by sagging walkways. The difference between them was that while Samal villages always had a gangway leading ashore, Bajau villages did not. There were other signs, too, of a culture becoming less isolate.

'Paganism' and the boat-dwelling habit have always been identifying marks of the Sulu Bajau. With the full acceptance of Islam and the abandonment of the nomadic boat-life, these sea folk will

---

[1] For this and many details of Bajau life I am indebted to Dr. Saladin S. Teo, a native of Laminusa Island, Siasi. In addition to being a superintendent for schools in Sulu, he has made a particularly sympathetic study of the Bajau over many years. His book *The Lifestyle of the Badjaos* (Manila, CEU, 1989) is a useful addition to the literature, but even more valuable to me was his friendly and courteous company on visits to Siasi and Laminusa.

cease to exist as a 'pagan' outcast people, and become amalgamated into the general Muslim Samal population of Sulu. Probably within another decade full-time boat-dwellers will disappear completely from the Sulu waters.[1]

This was what Nimmo wrote a quarter of a century ago, and since then the drift to land has become almost total, except for a few isolated cases. Yet his prediction of 'amalgamation' has certainly not come about, if this means the adoption of land-based social habits and values. To this day the whole problem of how the Bajau can be integrated remains unsolved. They are mostly unconvinced by the idea of education, so are often unwilling to send their children to school. Nor do they seem keen to learn new skills. And as for taking part in any social or political activity, it has proved almost impossible to interest them. They suffer, in short, from an admirable lack of ambition. Their relationship with the sea is so strong they give the impression of being only flimsily attached to land, and might leave again tomorrow if conditions became any worse. Maybe the sea itself is by the way; perhaps what they have in their blood is a nomadic indifference to roots. This might explain their amiable remoteness, their strange innocence. Since they have never owned property ashore they have always remained free of contaminating land squabbles, battles with landlords and developers, crippling rents and tribal annexations. At the first sign of trouble an entire settlement of Bajau may be discovered to have left overnight, in silence, their abandoned huts creaking slightly in the tide and their low craft already invisible over the horizon.

What has done most in recent years to change the Bajau's way of life is violence. No anthropologist writing in the 1960s foresaw that persecution would increasingly drive them ashore

[1] H. Arlo Nimmo, 'Reflections on Bajau History,' *Philippine Studies* 16, no. 1 (1968).

and that the shore with its press and clutter of people, its social cross-currents and complexities would prove a very mixed blessing. Nobody guessed they might have to inhabit a strange no man's land, an intertidal zone. But then, nobody realised to what an extent the Sulu archipelago would become a battleground. In 1974 Ferdinand Marcos sent in the armed forces of the Philippines against the MNLF. In the fighting of early February that year, most of the town of Jolo was destroyed and its population forced to take to the hills. Henceforth, the best that reigned in Sulu was armed stalemate broken by violent guerrilla and military engagements, until today's state of undeclared anarchy was reached. The great influx of weapons into the area, together with financial support for the MNLF from abroad as well as the money brought by lucrative smuggling and trade links with Malaysia, meant that the dominant tribe became more dominant still.

In recent years anyone has been able to acquire an M-16. If all else fails, one can easily bribe one's way into the army and acquire the weapon that way. M-16s are constantly 'lost' as soon as they are issued, and often the new recruit only waits until the weapon is in his hands before defecting. Armed with an M-16 it is simple to steal a boat. Anybody with an M-16 and a *bangka* can go straight into business on his own account as a pirate. It is a vilely dangerous living, to be sure, and the sharks must have grown very fat in the straits between the islands, but it is a way of life sanctioned by tradition and facilitated by the times.

The result for the unarmed, peaceable Bajau has been disaster. Whereas once they could fish at night using hurricane lamps and Coleman lanterns, they now dare not for fear of attracting pirates. A further disadvantage is that the pirate craft frequently have engines powerful enough to outrun the Philippine navy and coastguard patrols and whereas once the Bajau might have used superior seamanship to avoid trouble they are now helpless. So if anything has reconciled some of them to land and its unfamiliar ways it is the need to defend themselves. It is not Islam, nor free education by the Oblate Fathers, nor offers of health care, nor any amount of blandishments and promises which have changed the Bajau's horizon in Sulu. It is violence, and the necessity of earning enough to buy an M-16 and an engine in order to counter it.

Presumably, nomadism—whether of Bedouin or Eskimos or sea gypsies—is everywhere in decline. In order to survive, nomads need large tracts of unoccupied territory where there is no serious competition for their source of food, and such areas must be growing fewer. Besides, modern governments increasingly dislike 'floating populations' who seem ignorant of their control and who drift uncaringly across their borders and frontiers as if they did not exist. All centralisation is a threat to the periphery, and minor tribes which fall outside even the periphery tend to become fair game. It is a short step from being a minority to becoming marginal and then officially outlawed.

Even under average conditions, nomadic life is harsh, while a single stroke of ill fortune such as drought, epidemic, civil war, an oil spill or volcanic eruption can bring a people to the edge of extinction. To sentimentalise nomads is not a patronage they need. With an Armalite at last in their hands and deep memories of catalogues of injustice, they do not necessarily behave better than anyone else. What they retain, though, is priceless: a genuine remnant of the knowledge which has served the various species of *Homo* throughout his history. This knowledge is already lost to industrialised man and in this present century will be lost to the human race for good. It is a particular way of being in a landscape, of coexisting with ocean and land which takes account of minutiae we no longer know how to observe and maybe now cannot see at all.

There is a link between nomads and pirates and even smugglers. It has partly to do with living in a world beyond boundaries, but also with a detailed knowledge of that world which goes beyond mere geography. Pirates are simply seagoing versions of highway men or brigands; each calculates that his knowledge of the locality will be superior to that of any forces officialdom sends out to capture him. But in between moments of intense danger and excitement must be stretches of considerable solitude, and some sea pirates must themselves have a near-nomadic existence. After all, piracy need only be a sideline. At its lowest level, such as that which has all but driven the Bajau to land, it is a matter of rat-poor fishermen preying on other rat-poor fishermen for the simplest things, like a day's catch or a dugout boat. I am sure that half the wanderers who landed on 'Tiwarik' when I was there were neither particularly fishermen nor pirates nor smugglers but all three as occasion demanded. They might be best described as opportunistic nomads, and what characterised them all was that they were highly self-sufficient. It was not a luxurious life they led, but they were utterly at home in it. First and last, they were born boat people. All had that adhesive agility common to those who grow up barefoot on very small craft. Most had the tawny streaks in their hair, the bleached expression and frown lines of those who have squinted constantly at glaring horizons. All were skilled with dynamite, hook and line or woven fish traps. None was scared of man or beast but they were truly frightened of *mumu*, sea spirits and omens.

In its immense navigational complexity and its lavish range of hiding places, a tropical archipelago is ideal pirate territory, and piracy has been established for centuries in insular South-East Asia. Some pirates achieved fame and most Filipinos know the story of Lim Ah Hong, the Chinese pirate who in 1574 even raided Manila itself and came close to unseating the fledgling Spanish colonial administration. His name lives on, less for nationalistic reasons than because of a vast treasure he allegedly hid and which has been looked for ever since. (A treasure is, of course, any proper pirate's true legacy.) Down in Sulu, in Borneo and the East Indies, piracy always flourished well. This was partly because it was Muslim territory, with a complex assortment of fiefdoms and sultanates never brought under full control by any colonial power. When various Sulu potentates made alliances with their counterparts in Mindanao, the entire Philippine archipelago became prey to Islamic pirate junks. The more regular the colonists' shipping and trading became, the better the pickings, until by the nineteenth century piracy had reached epidemic proportions. 'From Mindanao to Sumatra, countless White travellers recorded their fears of, and warnings about, the savage marauders of the archipelago who thrived on massacre, violation and rapine.'[1] In 1830 Stamford Raffles himself had found 'no vessel safe, no flag respected'.

Today's predominantly Tausug descendants of those pirates who infested Sulu are merely carrying on a long ingrained tradition. Naturally, piracy can hardly thrive without victims, and in default of galleons carrying Spanish gold from Mexico there are interisland launches carrying people with wallets and cargoes of goods for Chinese traders. There are also the boats which run regularly between Jolo and Labuan Island in Malaysia, taking advantage of barter trade agreements under which copra and handicrafts are swapped for electronic goods, textiles and canned food. It may sound small-time, but each round trip can be worth up to $150,000, and certainly those concerned take it seriously enough. At the very end of January 1991, pirates killed twelve Sulu barter traders in a single concerted raid. As for smuggling, there is a brisk trade out of Sulu

[1] V. R. Savage, *Western Impressions of Nature and Landscape in Southeast Asia* (Singapore, 1984).

in marijuana, which also goes to Malaysia. This seemed unlikely enough, given that country's famously draconian penalties for drug dealing; but as I was succinctly told, 'Malaysia's a big place.' In return, 'blue seal' American cigarettes are smuggled back and are on open sale throughout Sulu and Zamboanga.

These are not romantic businesses to be engaged in, and certainly not to fall foul of. The treatment of the Vietnamese 'boat people,' the refugees who fled Vietnam after 1975, was a case in point, and those victims who lived to testify to pirate attacks—often by well-equipped Thais—gave horrendous accounts. The earlier waves of refugees were largely Chinese middle classes from the Cholon district of Ho Chi Minh City, lately Saigon, and were often wealthy. They brought what concentrated valuables they could with them, usually gold, hidden about the overcrowded boats with great ingenuity, sometimes suspended by brass wire beneath the keel. Maybe the earlier pirates were satisfied by good hauls, but as time passed and the Chinese were replaced by ordinary Vietnamese political refugees their savageries began to be mentioned in the world's press. All in all, it was a far cry from the behaviour of the Bangladeshi pirates who in November 1989 were reported as singing their victims a little choral medley before asking them to turn over their valuables.[1]

Instances of horrible and immensely daring opportunism abound among the archipelagos of South-East Asia and are frequently evidence of a nomadic understanding of the sea. Less than a year after the *Doña Paz* disaster another over-crowded Sulpicio Lines vessel, the *Doña Marilyn*, sailed for Manila out of Cebu despite the coastguard's warnings of the imminent arrival of typhoon 'Unsang.' On the night of 26 October 1988 the *Doña Marilyn* sank while trying to shelter in the lee of Guiguitang and Manok-Manuk islands off the north

[1] BBC World Service, *Meridian*, 11 November 1989.

coast of Leyte. On this occasion, at least, there was land near-by. The seas were very heavy and many survivors who managed to swim in the right direction were pounded against the jagged offshore reefs and died there. And yet while dozens of brave Manok-Manuk islanders formed a human chain far out into the surf to pull exhausted swimmers in, other villagers who had heard the ship's radioed distress signals and had seen her lights had long since launched their flimsy *bangkas* and were far out in the storm, hauling survivors aboard, stripping them of any valuables, and throwing them back in. Had they been rescuers, their courage could scarcely have been over-praised. Yet as plunderers their bravery was actually no less. They displayed a true piratical streak that night, amoral and enterprising.[1]

To make a living from smuggling, as from piracy, one needs to know the territory with a precise and local eye. This must be true whether on land or sea. In regions where particular trade routes run or particular economies have grown up, smuggling activities will develop their own skills, lore and traditions. The bootleggers of West Virginia who ran illicit corn liquor through the Allegheny Mountains developed driving and engineering skills for outrunning the law which in turn nourished

[1] To forestall any moral posturing we should remember the wreckers who within living memory were still active around Britain's shores. Their behaviour was arguably even worse since they were occasionally known to lure ships onto rocks by means of false lights. (For an outstanding recent account see Bella Bathurst, *The Wreckers*, 2005.) It is also salutary to go back sixty years or so to when London was mostly behaving gallantly during the Blitz. On the night of 8 March 1941, two bombs hit the crowded Café de Paris beneath the Rialto cinema in Coventry Street. One exploded, killing thirty-three outright and wounding sixty. In the semi-darkness, choking fumes, dust and general carnage, two men posing as members of a rescue squad went round calmly removing rings from the fingers of the dead and unconscious and turning out handbags. In fact, there were organised gangs who had an elaborate telephone network keeping them up to date on where bombs had fallen and which places might present the best opportunities for looting.

the entire sport of American stock car racing. Mountain boys drove as soon as their feet could reach the pedals, and apart from learning a repertoire of tricks (such as the 'bootlegger's turn') they also acquired great sensitivity to details of road surface, weather conditions, and a car's balance and handling depending on how full the hidden tank of liquor was. The archipelagic people of South-East Asia have analogous skills, but they have others as well which make land-based versions look coarse and two-dimensional. Above all they are prodigious navigators.

The Bajaus' ability to go back to a particular patch of ocean without reference to land seems uncanny. Stories are told of their being able to sail unerringly to a single lobster pot on an overcast night out of sight of land. I have never seen this, but certainly such things are habitually said about other sorts of nomads, whether Aborigines in Australia or Kababish camel herders on the fringes of north-western Sudan. They are peculiar to any people whose entire living depends on a knowledge of their natural surroundings and who are themselves largely bound into the ecology of the area. The Bajau's knowledge of the sea comes as much from living in it as off it and extends to its every aspect.

Anthropology has confirmed what was self-evident long before Thor Heyerdahl and his *Kon-Tiki* venture in the late 1940s. That is, that sea nomads have always been serious navigators. Fifteen hundred years ago the Polynesians were sailing around the Pacific in big catamarans using the stars, frigate birds, sea conditions, smells and their own stick maps to tell them where they were. (It would be interesting to know if these stick maps, whose intersections marked islands, shoals and currents with considerable accuracy, also marked *imaginary* islands which, over the centuries, gradually disappeared.) Recent research concludes that *Homo*, like many other species, does have a built-in sense of direction, no matter how atrophied it may have become from disuse.[1] Apart from navigation, though, a sea gypsy's knowledge of the ocean is scientific in its detail, yet his is very far from being a scientist's eye. For one thing, it tends to be holistic to a degree, whereas the impression given by most of the geophysicists aboard the *Farnella* was one of extreme specialisation.

The question finally asks itself: What order of knowledge do we stand to lose if and when such people as sea nomads finally abandon their way of life, and does it matter? Perhaps one can say with more than mere intuition that certain innate skills and faculties do atrophy through not being used, that an increasing reliance on electronics to mediate our apprehension of the world does lead to the loss of certain sensitivities and that to lose any sensitivity or awareness is limiting and unwise. Again, extreme examples are sometimes advanced in favour of retaining 'old methods.' In the case of navigation, for instance, it might be said that with increasing reliance on satellite-based positioning and guidance systems, the old skills of stellar navigation may no longer be taught even as a 'manual backup,' and will in time be lost altogether. 'What happens,' the argument runs, 'if something puts all electronic navigational systems out of commission at once? Suppose there is a massive solar flare whose radiation disrupts the GPS satellites?[2] Or one of those sudden reverses of Earth's magnetic polarity which would make it necessary to recalibrate all compasses? What then?'

---

[1] The evidence comes from laboratory experiments during which subjects' ability to distinguish North gradually improved, which 'suggests that orientation in humans is a latent sense, which in some people can be recalled very successfully after multiple challenge' (Mary Campion, *The Journal of Navigation*, 44, no. 1 (January 1991).

[2] Such flares are not uncommon. A few years ago, the electromagnetic energy of a solar storm induced currents in landlines which caused widespread power failures in Canada, blacking out entire cities. Recently, a similar surge of solar radiation was enough to slow several GPS satellites, altering their orbits and hence the accuracy of their information.

Of course, this is not quite the point, though there is a poignancy in watching the old and new technologies meet. In the early 1970s I found myself flying from Recife in Brazil to London Gatwick in a VC10 of British Caledonian. The aircraft was virtually empty. I was one of eleven passengers, and after the others had settled in for the night (they were mostly elderly) I was invited to spend as long as I liked in the cockpit. Such innocent, pre-terrorist days they were; casual in the economic sense, too, which is no doubt why the airline no longer exists. In the middle of the night the navigator, who had been getting radio fixes from Dakar and Cape Verde, stood up and opened a tiny hatch in the cockpit roof which he called the 'smoke ventilation hole.' This exposed a perspex bubble through which he shot the stars with a sextant. Today there are no more navigators in airline service, the last having flown on VC10s and Boeing 707s. The crew on the flight deck of a modern airliner consists only of the captain and the first officer. Neither has been trained to navigate by the stars. Nor has the cabin crew. If an aircraft is forced to ditch and its passengers and crew manage to haul themselves into the rubber dinghies they will not, unlike Captain Bligh and his fellow officers, be able to make a dogged landfall weeks later nor even, like downed World War II aircrews, know which direction to paddle in. All they can do is sit impotently bobbing up and down, waiting for rescue.

The point is not only what will happen if and when stellar navigation becomes a lost art but who, apart from astronomers will remain attentive to the heavens? And who apart from scientists will remain attentive to the sea? Even when it happens before our eyes it is hard enough to accept that species become extinct, that they always have and always will since without extinction there is no evolution. But the idea of bodies of *knowledge* becoming extinct seems quite as shocking, and it is difficult to see how it can be avoided when they are so inseparable a part of a rare and specialised way of life. It is too late now to save many a tribe—of Amazonian Indians, for example—who might have spared us years of suffering and expensive research had they been consulted in time about the medicinal properties of the plants they knew best. (This, of course, is the utilitarian approach to conservation.) Maybe after all, bodies of knowledge peculiar to a tribe should, like species, be allowed to become extinct once circumstances have changed and they can no longer adapt themselves.

Apart from rebelling instinctively against it, it is not an easy argument to counter. If in fifty years' time most Bajau are stockbrokers, what will the sea be to them except somewhere for family outings and expensive water sports? Of what use to future generations their present intricate understanding of the ocean? If there is a scientific rather than a sentimental answer it might be one analogous to that which sees the paramount importance of maintaining the diversity of species, of the gene pool. The more the world becomes politically, economically and culturally centralised, the more homogenised its ways of living, so the dangers of sameness become apparent. To take a notorious example, the EU regulations restricting the varieties of seeds permitted for sale within the union have for years been viewed as potentially disastrous by botanists and agronomists. A real threat is concealed in the preferencing of a handful of crop varieties chosen only according to marketplace (mainly visual) criteria. Once the genetic bank is depleted the chances of calamity caused by a single unexpected virus or pest become much greater. When in the nineteenth century the Irish potato crop was lost, creating mass famine and emigration to the New World, the potatoes were almost entirely of a single strain, uniformly susceptible to blight. In future, no amount of genetic juggling or selective pesticides will be as effective as growing the widest possible variety of fruit and vegetables, keeping unfashionable strains alive even if the immediate benefits are not obvious.

A consumer-based cultural uniformity is still some way off but is already advanced enough for certain grim futures to be imagined. At the same time, utterly various ways of experiencing the planet still do survive, though tenuously and in scattered fashion. The Bajau looks up, and the sun crossing the sky tells him any number of things, among them his place, his time, and how the sea creatures on which his living depends will be behaving. In another world entirely, one spanned with satellites and a global money market, the sun is just a noun, a hot and dazzling object rising with the Nikkei and setting with the Dow-Jones.

II

Poems to Tecla

From time to time, one notices around offshore corals, and sometimes even far out to sea, small insects skating on the water's surface. Usually they are to be seen in flotillas which mill frantically at the advance of a hand, but lone specimens can be found. They are water striders of the genus *Halobates* ('salt treader'), the only insects out of roughly a million described species to inhabit the open sea. There are thirty-seven coastal and five oceanic species of *Halobates*, and relatively little is known about their biology. Unlike most intertidal animals they have colonised the sea from the land, and face a seemingly hostile environment. They cannot fly; they cannot dive; and to watch them skimming the surface one would imagine they would be crushed by an eddy, let alone by a wave.

The more one examines them, lying in the water or uncomfortably draped over a boat's sharp prow, the more one wonders. It is known they eat the debris of smaller insects: midge corpses, and so on. It is known some of them lay their eggs on the blades of sea grasses and coralline algae. The mating behaviour of a couple of species has been studied, as well as their way of avoiding predators.[1] But much remains mysterious, not least how a land insect has found enough in its favour on the sea's surface to make its home there. If such a living can be made, why is *Halobates* the only known insect to have dis-

---

[1] See papers by W. A. Foster and J. E. Traherne, Department of Zoology, Cambridge.

covered it? And, infinitesimal speck on a liquid desert that it is, how does it ever find the tiny fresh corpses on which it feeds? What is its relationship with salt, and how does its physiology handle the salt economy? What happens to it in the torrential downpours which smash the sea's surface into foam and can put a drinkable layer of fresh water on top?

These are the sorts of questions which time and zoology will no doubt answer. In the mean time, one is left with constant surprise at how well forms of life adapt to conditions that seem impossibly harsh or daunting. The dislikeable neologism 'extremophiles' has recently been attached to them, though to describe an animal as a lover of extremes is an obvious anthropomorphism. Varieties of life colonise any place or set of conditions that will support them; they do not rate them in terms of human comfort. It was precisely these sorts of argument which helped the 'azoic' theory last as long as it did. Since then, of course, there have been abundant discoveries of life forms which have adapted to extreme conditions, including bacteria in active volcanoes; hydrothermal vent communities of crabs, worms and fish living next to 'black smokers' on the deep seabed with metabolisms attuned to near-boiling temperatures; lichens and mosses in the Antarctic; desert snails and extraordinary plants like *Welwitschia mirabilis* whose two curling, strap-like leaves winnow so much life out of blazing heat and nightly fogs in the Namib Desert that single plants may live to be over 1,000 years old.

'Extremophilic' even seems a not inaccurate way of describing certain members of the genus *Homo*. It is difficult to see a culture like that of the Eskimos or Bedouin as founded on timidity. Where nomads are concerned, it is clear there are certain cultures, as there are individuals, to whom wandering is a necessary part of the economy of living, not merely of survival. Just as people may grow up loving the most nondescript homes and surroundings, so cultures develop which are deeply attached to apparently unpropitious landscapes and conditions. 'Attached to' implies both love and dependence, in which it is not possible to distinguish between a living and the place where it is lived. Deserts and oceans, which to an outsider seem to share great similarities in that both appear virtually featureless and both are life-threatening without specialised knowledge, are places which transcend their own conditions to the point where some people consider them spiritual entities. To wring a living—still more a livelihood—out of them requires skill and courage but also love. In this case, a word like 'extremophilia' seems, if still graceless, quite appropriate. It is beyond understanding why governments and their agents should imagine that cultures which have taken millennia to attune themselves to such ways of life might cheerfully renounce them in as little as a generation. The world is dotted with groups of demoralised tribespeople, drunk in shacks and shanties at the margins of the societies that have disinherited them. Exasperated and not always unsympathetic officials complain about the inertia and fecklessness of Aborigines or American Indians, how pathetically they connive at their own degradation. Similar officials throw up their hands over the Bajau, baffled by their lack of interest in education or modern health care. But why would a Bajau wish to become assimilated or strengthen his ties with land when his entire history is based on the knowledge that in the long run life on land means nothing but trouble? The sea is not something he can turn his back on to order. To assume he can, or would want to, betrays the origins of the social science that expects it. Western, and particularly American, society thinks little of moving hundreds of miles to take up a new job; but to equate nomadism with mobility is a gross mistake.

When presenting *Homo* with an environment for living, the ocean strips away inessentials other than skill, knowledge and attention. It is a life which requires its own intensity and exacts its own discipline. Many people who are not tribal nomads by birth live lives as remote from ordinary society as that of the wandering albatross. This bird, electronically tagged, has been recorded as making single flights equivalent to the distance of Australia to Britain before returning to a particular island.[1] (In view of a life spent mostly on the wing it would be too cosy to call this place 'home'.) Among their many human counterparts would be the seamen who spend three or more years continuously afloat. One sunny afternoon on board *Farnella* I watched crewmen painting the deck around pink- and-white geologists as they sunbathed. Middle-aged men, they swapped a few Hull syllables with each other and painted as carefully as if the ship were a house they had clubbed together and bought with a mortgage. They paid no attention to the scientists, not even to one wearing a bikini and no top. They gave off an aura of austere contentment, as if they were pleased the *Farnella* was going nowhere while doing so along the most precisely navigated paths. In this way they combined seagoing professionalism with perpetual non-arrival. How trim the ship was! At three in the morning one would come across a man humming to himself, kneeling with a scrubbing brush in a lavatory in a cloud of bleach vapour. Val did not think of himself as one of that sort, however, since he admitted to being at sea just for the money. He had lost his woman and sold his house and now wanted to rebuild his finances.

'How else would I make £200 a week?,' he wanted to know. 'But it's no fun, I can tell you. Be honest, I loathe the sea. Why won't the bugger keep still? It doesn't seem to bother them. They're a funny lot. Some are here because they've got broken hearts. Ah, didn't know *that*, did you? True, though. Some because they've got no other home. Really, they're only happy at sea. Beats me. They dread going back, and that's true. A couple of days and they don't know what to do.

[1] *New Scientist*, 1759 (9 March 1991), p. 55.

'It's obvious why they're here, isn't it? Course, it's to avoid responsibility. No mortgage, no insurance, no tax, no car, no electricity bills, no gas bills. Free board and lodging. Those deck hands specially. Been at it since they left school, if ever they went, that is. They don't know any other life. Very ignorant. Very narrow minds, if you ask my opinion, though I don't mean that as criticism. The sea's all they know. They're mostly terrible people when they're ashore. They can't handle it, so they just get pissed. You never see them drunk on board, do you? Bit tiddly, maybe, but never falling-down drunk like they get ashore. No, they must really like the life. I've caught one or two of them sometimes, up on deck, just staring at nothing like a normal person'd watch telly. Mesmerised.'

All ocean drifters, 'salt treaders,' lone yachtsmen, mystics, island-hoppers, wanderers and hermits have a degree of impatience with, and ignorance of, the greater world. A further characteristic of nomads, which many of them share, is an absolute vagueness about geography combined with a precise knowledge of orientation. There is a sense in which no beautifully drawn chart can be made to coincide at any point with the inner maps they carry with them. The two do not relate to one another. I have met tribespeople deep in the western Egyptian desert who had no remote idea whether they were in Egypt or Libya. Nor, so far as they knew, did they hold any particular citizenship. The distinctions they made were linguistic and tribal, and the elaborate kinship tables each carried in his head amounted to maps. I would not be surprised to learn that some Eskimos are much the same, and while always knowing exactly where they are may not know if others happen to call that place 'Greenland,' 'Canada' or 'Russia'.

There is something irresistible about this since it affirms the ancient homogeneity of land or ocean, a unity of human experience that transcends temporary political boundaries. As for the seas, those vast tracts, seven-tenths of the planet's surface, of course they are mysterious and haunting. Time and again they draw people back to them: mad mariners, adventurers, solitaries, very often misfits on land who are transformed once they are afloat. Their keels cut tiny scratches on the face of an abyss of creatures and terrains which mostly will never be seen by human eye. Something satisfactory wells up from this deep and nourishes them.

There are extremophiles everywhere, and the adaptations they make to solitariness or small groups are various and strange. By no means all men forced by a sudden change of weather to spend five months' isolation in an Antarctic research station turn out to be either raving or overjoyed when the relief team arrives with thunderous bonhomie and promises of an unbroken year's leave. I was told of one of the last victims of *le bagne* to be released from prison on the Îles du Salut in 1954.[1] He had been there twenty-two years and had benefited from 'the cucumber solution' until his fellow convict and lover had died some years previously. The freed man, about to be repatriated to Grenoble, appeared unwilling to leave the island because 'he did not know how Tecla would manage without him.' Indeed, he begged to remain in Cayenne and to be allowed to go back to his prison island as a caretaker. This was refused. Sadly, he permitted himself to be dressed in a cheap suit and put aboard a steamer. He took with him a manuscript, 'very illegible, a poem of many thousands of lines, all written in pencil on different scraps of paper. I saw it myself, *monsieur*. It was a poem to Tecla, *about* Tecla, and for all I know *by* Tecla. And who was Tecla, you ask? Tecla, *monsieur*, was a gull with one leg. His companion of six years, or so he asked us to believe.'

A good example of a highly specialised social group living in extreme circumstances is that of seventeenth-century English

---

[1] My informant was the curator of the museum in Cayenne, c. 1972.

pirates in the Caribbean. It was described by a scholar who, in default of extensive documentation (for they were not great diarists) elegantly deduced by a series of negative proofs how their lives had to have been. His thesis is that these buccaneers were practically all homosexual and that their piratical activities were sustained by their sexual relationships, much as the Spartans' valour was. He cites as determining factors the generally lenient prevailing attitudes to homosexuality in England at the time and the way in which apprentices were drawn or press-ganged extensively from the bands of boy vagrants ('great flockes of Chyldren') who roamed the country and whose group identification, for their own protection, was exclusive. Furthermore, the population of the British West Indian islands was then almost entirely male, an imbalance enhanced by transportation. 'The single certainty is that the only non-solitary sexual activities available to buccaneers for most of the years they spent in the Caribbean and for almost all of the time they were aboard ship were homosexual.'[1] Very few pirates ever married, it seems, and those who did were uniformly unlucky with their women (and vice versa, one would imagine). Blackbeard, William Dampier, Bartholomew Sharp and other pirate captains jealously guarded their favourite boys, while all aboard took advantage of a form of male bonding discreetly named *matelotage*. It is a great pity there is such a dearth of contemporary documents, of poems to Tecla even, although vivid details do emerge. Captain John Avery was known as Long Ben, 'not because of his height.' Add to all this the occasional bouts of derring-do and the frequent orgies of drinking when every soul aboard from captain to ten-year-old powder-monkey was stone drunk, and by contrast with the solidarity of outlaw life afloat, the prospect of 'straight' life (in both senses) ashore would likely have appeared dreary indeed. It is strange to think

[1] B. R. Burg, *Sodomy and the Perception of Evil: English Sea-Rovers in the Seventeenth Century Caribbean* (New York, 1983).

that, but for the lack of a few hundred women in the West Indies, piracy might have assumed quite different patterns or even have been suppressed entirely by privateers.

Nomads have a need to wander in a world they understand. Like *Halobates*, many tribes and individuals are very specifically adapted and cannot resist gross change. Take away their habitat, their rovings and solitudes, and they go to pieces. Because of the prevailing cultural pressures in Sulu the Bajau are mainly Islamicised and have become or are becoming Muslim. It is an appropriate religion because Islam contains a statutory requirement for pilgrimage. Perhaps since Islam itself evolved from nomadic cultures it made pilgrimage one of the four chief ritual and moral duties incumbent on all Muslims. In Christianity it is a tradition which has long since faded, except in occasional re-enactments. There is a psychological accuracy in this insistence that a proper life cannot be lived without pilgrimage, a journey, a great excursion and abandoning of town, village, hearth. Only in this way is the unsuspected majesty of the world revealed. Only by travelling in danger and discomfort along arduous forest paths, desert routes and sea lanes may a truth be approached.

What is it that a Bajau sees? In one sense, more than any oceanographer; in another, less. In that his vision is peculiar to his living conditions, it constructs a set of knowledge about the sea and the world which in the long run must be doomed, although it could be partly resurrected by anyone taking up that life. Much of this vision derives from a literal and daily view of what goes on under the sea. It is informed by the two major characteristics of the tropical waters he inhabits: clarity and phosphorescence. By day, the water's transparence declares its emptiness; at night it is filled with blazes of cold fire. This powerfully strikes even those who can talk about phytoplankton.

In the coastal towns and settlements of Sulu the outermost stilt huts of Samal and Bajau may stand one hundred yards off-shore, extending landwards in a dense maze of duckboards and walkways and pilings. Having landed on the rickety wharf it is often impossible to determine exactly where the shoreline is. Gradually the gangplanks between hut and hut sink lower and turn into paths made of embedded lumps of coral the size of babies' heads, damp causeways below floor level with the glint of water on either side. One can often suppose oneself well ashore and then glimpse sumplike pools beneath the huts littered with a floating debris of rotten thatch, splinters of bamboo, boatbuilders' wood shavings, plastic bags, bobbles of excrement, a dead kitten. One would expect these pools to be foetid, miasmic, black. Anywhere else they would be. Yet a muddy bottom is generally visible, often startlingly clear between the bits of flotsam. Perhaps this is simply because the daily tidal creep is just enough to prevent the water stagnating, although any rise and fall is imperceptible.

Rather, it is as though the seawater here contained a partic-ular natural ingredient, *transparency*, which is proof against the cloudy, the dank and the foul. One imagines the water so vibrant that everything in it is strongly polarised. Things are either abundantly alive or else they are whitened bone and shell. There is hardly time for any intermediate stage of decay and it darkens the water not at all. Away from the huts, out over the reefs and shoals, the water is a huge lens focusing light and embalming everything in purest blue glass.

We would say this extreme clarity actually testified to a deficit not of impurities but of nutrients in the waters of the tropics. These seas are relatively stable compared with the colder, stormier oceans. Hence in their euphotic zone the nutrients tend to sink straight down, and photosynthesis is cor-respondingly slower. The deep blue of tropic water is the sea's equivalent to the ochre colour of deserts and, likewise, signifies an absence. It is hard to believe that cold grey northern seas should be more nourishing than pellucid tropical oceans with their teeming reef life; yet it is so.

After dark, however, these apparently thin waters are rich with the lights they contain. One moonless night, I watched some Bajau, scared of lighting a lantern for fear of attracting attention, dive in total darkness for a lost saucepan. The pot had sunk in shallow water, only two or three fathoms, but it still appeared a hopeless task.

I have never seen phosphorescence as bright as on that night. Leaning over the edge of the *bangka* I could follow every move of the searchers below. Only, the whirligigs of sparks, the flashings and showers of cold fire were at depths which could not be determined. Just as the glints and refractions in the best opals can appear deeper than the thickness of the stone itself or else closer than its surface, so the divers' movements excited discharges of light which were either a few feet away or in a universe beyond. It was vertiginous to gaze down because the view was more what one normally expected to see overhead. On nights as dark as that, it is anyway hard to define the horizon, to separate black sky from black sea. Now it was as if the cosmological figures of Sagittarius and Orion had come to life in a firmament beneath the boat. Legendary men outlined in stars swam among clouds of dark matter, galaxies and nebulae swirled in the eddies of their passing. In that moment, I could have glimpsed a figure of the Bajau's own believing, the *sama sellang* or ocean gypsy who lives in the depths and sometimes leaves his kingdom at night to walk the beaches, a black and shining giant twelve feet tall. He is King of the Fishes, and any Bajau wanting to fish in his domain should have the courtesy and sense to ask his permission first or risk having his boat turn turtle without warning. It was while I was leaning over and watching that one of the outriggers inexplicably sheared off, tipping the *bangka* over and myself into the sea in a sheet of flame. In only a minute or two the Bajau had found their pot on the seabed by the light of their own hands. They emerged with it, laughing, and we all hauled ourselves from the ocean running with greenish fire.

Such are the nights when it is hard for a swimmer to resist heading downwards, trailing constellations in a fading dust, and simply go on swimming into the fathomless, sparkling spaces below. It does not feel suicidal in intention, nor like an attack of calenture.[1] Rather, it is more akin to something which might also have played a part in the Bajau's ancient renunciation of land: a momentary expression of yearning for oceanic roots.

---

[1] Calenture used to be defined as a tropical shipboard fever whose intolerable burning sometimes made sailors jump into the sea. If this seems over-determined, a secondary meaning has gradually been allowed to surface to explain the behaviour of individuals who, though feverless, may leap overboard without warning and vanish. In this sense, calenture becomes a species of mystical rapture, a yearning to blend with the infinite, and has been cited to explain baffling disappearances by lone yachtsmen like Donald Crowhurst and—more recently—the death of the newspaper proprietor, Robert Maxwell. If the condition could take a mass-hysterical form, it might even throw light on the enigmatic desertion of the Mary Celeste in 1872.

It is one of those days when everything we do feels final. The sea is stained with cloud and veined white with its own collapse. It is as if we were having to bid farewell to an entire set of knowledge, to a lifetime's habits. Nothing will come in their place, not epiphany, not even a proper death. Nomads, too, keep track of the seasons, the stately heeling of the stars. So an anxious internal creature (oh modern man!) is forever flicking over its left wrist in a gesture unknown before the invention of the wristwatch. This brisk flip of the knuckles encodes a world of power and—by implication—of powerlessness before a constant anxiety. No doubt there was once an equivalent gesture for a man of the world, perhaps a particular straying of the right hand as towards a sword's pommel, half unconscious, the weapon (no more than an article of dress) forever undrawn, just as the time remains unmemorised, somewhere between a fossil gesture and a social tic.

One need not wear a watch to feel an imaginary arm bend and a wrist turn. We glance at the sun. Even that looks final, seen through the haunted lens of unease. Something is coming. Something is on its way to break things up. We are vaguely prepared for it, unable to settle to anything, as on the day before a journey, knowing it is out of our hands. When it finally arrives it will happen to us, willy-nilly. On a day like this, look up at the sun and the clouds. Look at the sea. It is all written there.

Meanwhile, what of the swimmer who lost his boat at the beginning of this book and was left all alone with his panic in the middle of the ocean? That he is here to write the question means the sharks did not get him. Nor pirates, nor fishermen. The engine he heard was another figment. After a long, long time his sight cleared or light rays unkinked and there, no more than eighty yards away, sat the boat. It was solid and unmistakable. It had the air of never having moved, of being practically nailed to the sea, while the swimmer immediately felt his limbs achingly heavy as though he had been on a long and wilful excursion and had returned to his senses in the nick of time.

More mysterious still, regaining the boat was like putting on a pair of spectacles, for the low palm-fringed coast became visible exactly where it ought to have been. Ever since that day the swimmer has been unable to account for what happened other than by using a cryptic phrase such as that he fell out of one gaze and into another. He had made the visible a little too hard to see, even though his life depended on it.

Later, he decided the sensation had been less of being lost in the sea, or lost to the sea, than of the sea's *being lost to him*. He was surrounded by water which could have engulfed him; yet at the same time it was a sea which had receded in a way not immediately obvious, taking with it whatever was essential—knowledge, perhaps—for survival. For a long moment there was a boundary fixed around him, an exclusive zone of taint, while perhaps monsters did swim up unseen from the deep, sadly, to stare at a pair of tiny white limbs cycling high in their skies on the very edge of space. Even had they eaten him, the time of their dominion was past. Eventually the legs vanished and the swimmer made off, leaving silvery paddle pocks like fading footprints. The long subsequent journey, of thinking about the sea and the oceans, showed him he was treating them as something which had already been lost.

The gaze of the nineteenth-century explorers and oceanographers was emphatically not the same as our own. They still saw through the eyes of mariners, wary and respectful, as well as with the awed curiosity of intellectual men to whom the ocean was a vast leftover from Creation, a divine challenge which could be met only with the right degree of humility, bravery and methodical amassing of facts. Raw nature was often fearsome. Wild animals were mostly a menace, and even if killing them was not strictly necessary for survival, their deaths spoke of a proper relationship between man and nature, sanctioned by Scripture. 'It will not be denied, we presume, that animals were created for the use of mankind.'[1] In this clear light, landscapes, too, were a threat. Many mountains were first climbed in the nineteenth century as much to tame them as anything else. Nearly impenetrable areas of Africa were explored and named and dragged on to the map, into the domain of 'civilised' man (the native gaze of those already living there, and especially the nomadic gaze, did not count). Such was a colonist's view with acquisition on his mind. As for regions like the Antarctic, 'Great God,' exclaimed Scott, 'this is an awful place.'

But now, and mostly within the space of the past few decades, this view has vanished, and modern eyes see an utterly different, less awesome, world. Apart from the fact that nothing a modern oceanographer could discover in the ocean would be likely to precipitate a spiritual crisis and that technology has made 'mariners' in the old sense obsolete while removing a good deal of science from the domain of humility and bravery and dumping it instead into the frankly humdrum—apart from all this, we can now only ever look at the natural world with the anxiety of conservationists. It is all falling away, becoming lost to us. No longer are the wilds and unoccupied spaces of Earth things to be tackled and subjugated. Rather, they are to be cherished and protected because on them depends our survival. The view through our sunglasses and snow goggles and diving

[1] Editorial, *Edinburgh New Philosophical Journal*, 2 (1860), p. 283.

masks is suddenly of last things. The midnight chorus of a reef filling our ears as we clutch in darkness twenty feet down is the voice of the sea rehearsing its own eschatology.

Conrad dated the beginning of the end of the old sea from the building of the Suez Canal, completed in 1869.

> Then a great pall of smoke sent out by countless steamboats was spread over the restless mirror of the Infinite. The hand of the engineer tore down the veil of the terrible beauty in order that greedy and faithless landlubbers might pocket dividends. The mystery was destroyed . . . The sea of today is a used-up drudge, wrinkled and defaced by the churned-up wakes of brutal propellers, robbed of the enslaving charm of its vastness, stripped of its beauty, of its mystery and its promise.[1]

That was first published in 1896, and its jeremiad tone is similar to many other laments over industrialisation. Over a century later, only vestiges remain of the sea Conrad described, let alone of the one he mourned.

Yet even this is not quite the reason for the swimmer's conviction that the sea was lost to him. The oceans have long been, and will long be, subjected to ruthless exploitation and even, in places, to ruin. It is not really the sea which is in recession, though, but wildness itself. Wildness is everywhere but it can no longer be seen; and its apparent vanishing is a direct consequence of the new conservationism. 'The Wild' is nowadays a concept ringing with the overtones of patronage, of collections by schoolchildren on its behalf. The present generation is as much contaminated by its own reverential and placatory attitude as the older was by domination. There is something ignoble about it, compounded as it is of urban sentimentalism, virtuous concern and sheer panic at having irrevocably fouled the

---

[1] Joseph Conrad, *An Outcast of the Islands* (1896), beginning of Chapter 2.

nest while so comfortably lining it. Above all, the self-interest shows through. Luckily, there is a chasm properly and forever fixed between the non-human and the humanist biospheres, between wildness and *caring*. It is seldom visible to modern eyes. Virtue and the wild share no common universe.[1]

If the sea always was a rich source of melancholy, there is in any case a new melancholy to go with the new view. Conservation is only ever a rearguard action, fought from a position of loss. It is ultimately unwinnable, and not least because there are few recorded victories over population increase, as over the grander strategies of genetic behaviour such as the laws of demand, political expediency, sheer truancy and refusal to relinquish a standard of living once it has been attained. There can only be stalemates, holding actions and truces uneasily policed. A few affecting species will be saved, a few million hectares of forest, a few tribes of Indians; but the world will never return to how it was when this sentence was written, still less to how it was when reader and writer were born. This has always been true and will continue to be so. The mistake is to extend this sequence backwards in time and imagine it leads to a lost paradise. It is a safe bet that as soon

---

[1] It is a wishful belief on the part of many that in some mystical way animals can appreciate human social values such as 'goodness.' 'They know,' is the usual sage assurance, 'they can tell.' On the contrary, all evidence suggests that animals have not the least interest in morals, or else they are remarkably undiscriminating. There is an account in Browne and Tullett's biography, *Bernard Spilsbury*, of the murderer Patrick Mahon. In the 1920s, in a house on a desolate stretch of shingle near Eastbourne, Mahon cut up and rendered down in a cauldron the *disjecta membra* of Emily Kaye, the girl he had made pregnant. Mahon, 'like St.. Francis, whom he resembled in no other way, had a remarkable influence over animals. Those who like to think that animals know good people from bad will be distressed to learn that Mahon had only to whistle and birds came to him, and that dogs and cats deserted their masters and mistresses to follow him home.' One assumes the explanation was something like pheromones. Likewise, animals flocked to St. Francis not because he was a saint but because they happened to like the smell of his glands.

as the earliest proto-hominid could think, it invented a legend to account for its sense of loss.

But the *swimmer* . . .

The hypocrite swimmer has himself lost all interest in these arguments. He is intently reaching the shore in his little boat, paddling carefully among familiar corals, following the narrow channel in towards the beach. He has already forgotten his panic and is merely tired as from a long journey. Cool in the bilges lie half a dozen mackerel: two for supper, two to smoke for tomorrow and two to give away. He looks up as the prow grinds into the sand. There in the palms' ragged shade is his lopsided hut, there the tangle of thorn shrubs concealing a mahogany-coloured brackish lagoon, in the distance the spit of mangroves walking on water. He sees it all not through the eyes of an oceanographer, still less of a conservationist. Only in a nomad's or a wanderer's gaze is the sea not lost to him, nor any less wild. So affectionately does the scene bound towards him and leap into his eye that he knows this private way of looking reveals a landscape he must have inherited, or which was somehow fixed for him as a child, before he ever saw it for the first time.

For this is his ocean, and at last he knows he has always seen it thus, towards the end of afternoon: the great white clouds heaping themselves out of nothing against the blue, their tall reflections falling on a glassy sea whose tide lies stilled at low. Reef tops knobble the surface, the kelps and grasses float as rough brown patches among which the white clouds lie in fragments. Children stand a hundred yards out, up to their ankles, legs angular as wading birds, filling coconut shells and tins with winkles. They dabble among the white clouds. Clear voices drift ashore, tatters of a heedless present.

It is the moment of being aghast at the sad miracle of having condensed from nothing, of watching white clouds, of dispersing again. But how beautiful it is; and how pierced by it we always are as it leaps through us and, leaping, vanishes.

III

MARGINALIA:
Steamers/Streamers

T he Sulu archipelago is a good example of a place which must be reached by boat if it is ever to be seen. Only a boat, as opposed to an aircraft, will put the traveller within its coordinates. There is always some risk of attack by pirates or of going down in a vessel like the *Doña Marilyn,* and it is important to court that risk. Besides, the cramped, hot, vomity approach through a sea sprinkled to the horizon with small islands is the correct one. It is necessary to wake at dawn on a folding canvas deck bed jammed between its neighbours like a stretcher in a busy field hospital, face clammy with salt and dew and whipped by strands of a stranger's hair. Out of that fitful, blurred sleep, an island has emerged on a turquoise sea and those whose destination it is begin to stir, waking their children, pulling their belongings together. This slow, oneiric approach must be observed. No place ever quite survives the wrong landfall.

If one wished to formulate a First Law of Travel, it might be that the mode of travel determines the place reached. To take an example: the Korea one reaches by cycling from Hamburg (which could be done without ever once glimpsing the sea) is altogether a different place from the Korea reached by flying from Heathrow or Kennedy. The people are different, they speak a different language and have a far greater knowledge of bicycles. This principle holds good for Sulu, too, and is likewise dependent on adopting the right pace. All those passenger ships, more or less unseaworthy, which still ply the networks of archipelagic routes, are simply ferries, shuttling people around within a country. They make one lament the passing of international sea travel. The sundry boredoms and discomforts of passenger liners were as nothing compared to those of the aircraft which have replaced them. There is a new generation of cruise liners, it is true, but they scarcely touch the argument. Cruise liners are not going anywhere, so they function more or less as hotels, with the novelty that they float and move. They are as aimless as the pleasure they pursue, that classic wild-goose chase. As was memorably said in a radio programme about the hijacked *Achille Lauro*, however luxurious a modern cruise liner, 'it's really no more than a velvet-lined prison hulk.'

Just as modes of travel affect destinations, so do they change our farewells and the very nature of separation. Airport terminals swallow up friends and lovers in a way docks never did. The drabness of quays is not to be compared with the squalor of concourses. Even if we bother to wave foolishly from the terminal roof as an anonymous metal toy half a mile away disappears within seconds into grey overcast, we can never be sure it was the right one. In any case, part of the dispassion of these aerial bus journeys lies in their being so swift. It now takes effort and care for two people to be distant from one another by much more than twenty-four hours.

Ships, on the other hand, carry with them the solemnity of long separations, perhaps of lifetimes. Stately valedictions echoed through people's lives until forty or so years ago. Ships are individual as one of a fleet of Boeings never could be, even though an aircraft may carry a name (often more or less geared to the tourist age: *Loch Ness, Val d'Aosta,* etc.). A ship *is* her name, right to the bottom and beyond, connoting a moment of history as much as a vague locus on the seabed (*Titanic, Lusitania, Andrea Doria*). Aircraft, when they crash, shed their names along with their wings. They become 'ill-fated Flight

307'; or else Pan Am Flight 103 becomes 'Lockerbie,' an entire nexus of loss reduced to a point of impact.

The departure of a ship is slow, celebratory, mournful. It gives time to think and the proper space in which to let fall one's lesser salt into the greater below. Something of moment is happening, part of whose subtext is a fear or resentment of the sea as the agent of long absences, slow letters and terrible news. Whoever they are, down at the docks one windy afternoon—friends, lovers, siblings—they are already separated. There are those on the quay and those already on board, though both are watching. The ship is about to sail. Gangways are lowered, ropes cast off. Heavy nooses splash into the slot of oily scum between lorry-tyre fenders and iron cliff. Cries go up. The siren's blare, of such low frequency it shakes the stomach and jars loose fresh tears, sounds once, twice. Yet an illusion of unparting is preserved by the streamers, cheerful strips of paper sagging and twirling between the thousand pairs of widely separated hands.

Over the whole scene hovers loss looking for somewhere to settle. Is it in the already spoken goodbye? In the last touch of bodies? In the cries of the gulls? Or does it now pulse along that thin paper nerve? It parts; they part. Yet still they remain visible to each other while loss fills up the space opening between them, stretching out between ship and shore, between hull and headland, dot and smudge, before spreading across the face of the globe. But of course it had been there in the train on the way to the port. And before, in the careful packing of suitcases. And before that.

And after? Here again, air travel offers no comfort because its speed runs departure into arrival, leaching out their difference, blurring them into a hectic placelessness. We cannot tell what to think. It is too brutal, facing the ordeal of a dawn landing on a strange continent with the scents of leavetaking still in our clothes. We have walked the streets of an Asian city with the fresh scratches of a cat in Oxfordshire on our hands. Such confusions make unreal both cat and city and leave us feeling we can never properly come to grips with anything.

Travelling long distances by sea, on the other hand, gives us time. Travel is like death in that it requires mourning. The light melancholy of watching a coastline recede is a necessary observance. We join in with shipboard life just as soon as we wish, and not before. Otherwise we write in our cabin or spend hours watching the wake of our own passage. The caves sucked into the water's surface by the turning of invisible propellers—each subtly different, each marbling a dissipating track which stretches back, an elastic streamer—become hypnotic. They set us adrift on inward voyages where we barely have enough sarcastic energy left to stop ourselves seeing our frail barks upon the vasty deep as paradigmatic. Such time, such long hovering on the edge of banality, is powerfully restorative. By the time approaching land is announced we are free to be excited. Later, it seems to us that only by having breathed the salt air of loss for long enough are we able to make a properly carefree disembarkation. We have adjusted. Our biological clocks are reset, our homoiothermal balance has altered with the latitude, our internal maps—whose every nautical mile has been *felt* as travelled—make sense. Behind us the ocean is criss-crossed with thousands upon thousands of multicoloured streamers, a planet festooned with farewells.

8. Sea Burial
(Thalassotaphia)

'Burial of de Soto'

I

D o you remember this?
A broken engine; the hours' becalming; an empty
ocean still as a lake of mercury. It was soundless to the
horizon and our small noises placed us at the centre of the universe, unique in our activity. From time to time a spanner
clinked, a bare foot bumped a thwart. From the vinyl-scented
shade of a rigged tarpaulin we watched the foot's tiny ripples
become visible only as they left the *Medevina*'s shadow, trembling outwards, as if the shadow's edge were the actual hull,
our whole craft insubstantial, no more than an airy nothing
which had briefly come between sun and sea. The mutter of
voices – 'Try this. It's rusty. The gasket's ruined.' – the rasp and
flare of a match, the incense of a cigarette. The small splash of
a handful of waste. And then, emerging from the shadow into
gorgeous colour like the tip of a kingfisher's wing, an iridescent
oil-stain flashing its molecules, splitting the spectrum and
creeping out across the water. Do you remember how dazzling
it was? That spreading puddle of hues in a still world of primary blue? Greens and purples, golds and pinks, rubies and
violets, forming and re-forming, pooling and glittering minutely
so the fascinated mind drew ever closer to its surface and fell
into a microscope's gaze such that the twinkle and sputter of
evaporation almost became visible, the spirituous fractions
boiling off in order of their volatility.

The sun climbed and remained stuck at the top of the sky.
Sometimes we stood up or wandered aft to peer at the disman-

tled carburettor. The parts were black with oil and rust: deformed, even—corroded artefacts turned up by a plough rather than precision-engineered components. Our shifting weight as we moved about the narrow boat made one bamboo outrigger gently dent the water, the other rise and shed a line of droplets. The brilliant oil-stain fractured. Feathers and petals broke off, some drifting perversely back into our shadow and winking out. A flotilla of melding islets moved into the glare beyond the outrigger, sending back scorching chromatic flashes. The hours passed. Fish and rice to eat, the bowls washed over the side and fat replenishing the film until—do you remember? – a peacock sheen surrounded our soundless universe, marbled and swirled and striated. In a halo of specious glory our little boat sat and baked, breathing out its rainbowed anima.

No, you will not have forgotten; not in the light of what was to happen later, not in the particular isolating light in which individual events were picked out with such intensity, casting profound yet insubstantial shadows.

Now and then a fish rose, but languidly, as if its head had difficulty breaking the surface tension of the mercury. So thick was the water's skin, the animal's head should have been ringed by a downward-curving meniscus, a pucker. Each occasion seemed a dramatic event, and several pairs of eyes would remain long afterwards on the resealed hole where snout or fin had protruded for an instant. At length, a flying fish broke completely out, tail whirring the water's surface like the propeller of a planing speedboat and leaving a straight scar of irregular dashes for thirty metres before vanishing into an invisible notch. This was a fishing boat, after all, and at last someone aroused himself out of his lethargy enough to bait a hook and drop it into the rectangle of water enclosed by an outrigger. It was as much for something to do as a gesture of habit.

What is it that makes even a fisherman forget the bulk of the world underlying his tiny craft? He moves too much in air, ethereal, reading a visible text in light's plain version. He waits and ponders and is becalmed in the eye of the sun, turning over diurnal and terrestrial thoughts. He tinkers with carburettors. The illusory line separating air from water, dividing the lighter swirl of molecules from the denser, merely compounds the fiction of two worlds dwelling apart, the one inimical to the other. Yet what could be better proof of their radical contiguity than the gallant life force pervading both? Not in a mystical sense, either; but because life originated from below, some of it adapting to permanent exile and some of it staying put. We are colonials. What we have in common with our ancestors is the sea and not the air. How could a fisherman daily pull from one medium into the other his forebears' struggling descendants without having a mind to that world of relatives beneath him? Every day the children of his archaic self observe his keel's sliding footstep as it treads their upper atmosphere, crossing their heavens with its contrail. They underlie him as a half-forgotten nation underlies its vast diaspora. One day this has to be acknowledged; one day he must return to his roots among the vents and plumes, the ancient dark upwellings, his cradle's gyre and nurturing oblivion. Maybe only the exceptional fisher does not simply make daily raids across the frontier, poking spears and hooks and nets into its first few metres. Maybe the exceptional fishers are those who sense their own roots extending deep into dissolution and can feel the freedom and excitement of the abyss.

It was mid-afternoon before the engine started. Our companions murmured their relief. We had lost six hours. The blocks of ice in the insulated chests would be that much smaller. Nobody had opened the lids to look, for fear of accelerating the process. Now we could spend fewer than four hours at the lobster beds, even if the motor kept going and took us there before midnight. You will not have forgotten how precarious these trips were, ninety miles each way. If all went well we could fill half the chests with lobsters, having transferred the remaining ice to the other half to pack around fish. Any serious delay on the way back might mean the fish were lost. And this quite apart from the usual risks of dodging coastguard cutters alert for illegal fishermen. Edgy enterprise, but grounded in comradeship and marked, as always, by heightened senses.

How vivid, still, are the seagoing smells? Oily bilges, fish entrails, a freshly lit cigarette drawn through salt paper? And at night if you were not diving, the compressor's exhaust fumes, its lethal monoxides, barking and blattering our darkened boat's position for anyone to hear. But a shift of wind might gently lay its hand on a cheek and turn your head like a weathervane, pointing your nostrils into the smell of unseen land: forest and rot and copra, jasmine, mimosa and ylang-ylang. And you may have thought of the strangeness of it, sitting there in night's scented cocoon, propped up by nails and timber in the middle of water while men you knew like brothers worked away in the fish mines far beneath the boat, their dim torchlight opening up fugitive seams and corridors. Their wooden goggles and floating hair.

And behind everything, the economics: the tanks of fuel we carried set against the miles we needed to cover; the ice matched with that and the anticipated size of the catch; the future price of fish in the market half a week away; profits eaten up by repairs. The fisherman's immemorial equations, whose hidden term is time. Tides and currents, nightfall and winds, the rising of the moon: all determine whether he is too early, too late, or punctual for the flow of fish he is banking on at his journey's end.

So we were already late; but the captain said, 'Push on. Push on, *go!*' (pointing with jutting lips to the speckless ocean's rim). To turn and head for home was too safe, too undisguised a loss of fuel and effort. Follow the sun, now curving a south-westering course. In this way we reached the moment when our prow was aimed directly at where, in one corner of the day-long featureless sky, evening clouds were beginning to form a range of wool mountains above an invisible coast. It was still too early for pinkish colours. The peaks were white and blue and white, undershadowed with pearl, glittering snowfields which convection was thrusting into pinnacles and spires even as we watched, transforming the alps into the skyline of an im-possible town. There, among amorphous towers and eroding castles, an insane architect was busy opening up sly passages, glimpses of alleys, vistas of tall windows and bent porticoes whose very act of moulding held out the constant promise of disclosure. Glancing away for a moment and then looking back, the eye would catch these apertures at the tantalising instant of being sealed by vapour, blocked by soundless landslides, erased for ever. Even as the eye scrambled back down crumbling stairways leading nowhere, the great gates were slamming all around in silence. And always, as they closed or melted they left behind the faintest air of having successfully denied, of having withheld a view of some innermost chamber or secret courtyard now buried deep within palisades.

Meanwhile, the mercury ocean across which we had been heading was gone. In its place a glassy violaceous swell reflected this soft metropolis so the one image leaned above the other, the lower scarcely distorted. And it was on this, dead ahead, you will remember we first saw the tiny black insect of another boat. The only visible object on the ocean's face, it lay in our path with the punctuality of an omen. Without needing to alter course by a degree we gradually overhauled the lone fisherman as he sat in the celestial city's mirrored thoroughfares.

The captain eased the nylon string which held the throttle so the engine note fell to a mutter and the *Medevina* lost way. With the rudder slightly over we began a gentle curve which took us within hailing distance of the fisherman but not so close that our wake would swamp him. As to what we already thought, can you remember that? Even though we had passed many identical craft on previous expeditions, can you remember if you had taken in the absence of a mast and rice-sack sail, or the fact that the boat's smallness and the man's position in it made it obvious there could be no engine? Even as we came abreast thirty yards off, had you appreciated that nobody under paddle-power alone would allow himself to drift so far out to sea that no land was visible at any point of the compass? A stocky man, the boat's occupant was wedged comfortably low on the bottom boards, watching our approach from beneath the brim of a hat shaped like a straw lampshade. A fishing line was stretched the freeboard's short distance between one hand and the water.

'Oy, *paré!*' called our captain, allowing us to laze across his bows, then around to the other side. The fisherman's face did not turn, however, and something about his posture had forestalled any jocular shouts of '*Hoy gising!* Wake up there!' The captain only said for all of us, softly, '*Yari na. Patay na.*' And so we completed our circle, staring at the dead man, the water at our stern dimpling and crawling above the scarcely revolving screw. There still was no breath of wind; but possibly even at a saunter our own larger vessel with its spread awning leaned against enough air lazily to displace it, for by the time we had gone entirely around the fisherman, a waft of corruption had reached us. There, in the early evening light amid the shattered debris of clouds, he sat and exhaled the gases of his own corporeal breakdown.

II

It is well known in these parts that fish choose not to speak in order to risk nothing worse at men's hands. Being wrenched from the depths into thin and bitter light to drown slowly in air is bad, but not bad enough to merit speech. Suffering put into words merely takes up an additional burden of pathos. Even being laid alive on coals may induce them to writhe but to utter only occasional squeaks of steam. Hereabouts, people say fish are wise not to give tongue and express a reasonable protest at such treatment. If they did, they would lose the last refuge of dignity which stoicism confers: the right to die unobserved by the faces poised over them, fanning and chatting and waiting for their supper. To enter a mild reproach would excite their captors' amazement and curiosity so that henceforth the fishermen would bring their catches to the point of speech in delighted competition to see whose was the wittiest, whose the most touching. Sensibly, the fish remain silent lest men either patronise them or redouble their torture. Nor do they mourn either themselves or their fellows. By such means, it is said, they keep their spirits intact.

As the *Medevina* slowly circled the fisherman, we appreciated how in the end he had learned from a lifelong identification with his prey. His own muteness surrounded and exalted him. He was sitting deep in a tiny boat which was anyway lower than ours, his buttocks wedged below water level, in a sense already partly submerged. Objects found in mid-ocean sometimes look larger than they are, as though defiant at being so consummatcly dwarfed. This fisherman's silence girt him about like the ring of ornamental chains surrounding a great monument in a public park. And as the chains mark off a space where the horizontal becomes the vertical, delineating a change in the mode of looking as well as a boundary preventing intrusion, this man's towering privacy and immobility kept us at a distance until we could work out a way to approach him. He had a paralysing effect. The very fact that there was nothing to be done left us with everything to decide. (In these parts, then, the spirits of stoical fish rise up from their driftwood pyres and eddy about the eaters, listening to the mirth and chatter, trying to understand their executioners. Some float off down the beach to the sea, where they creep out across the water and wait for the time when they can be reabsorbed. This needs complete darkness, the black of an overcast and moonless night. If, as so often happens, the sea is bright with the reflected light of galaxies, the fish-spirits hang sadly above the surface, unable to return to their world. With the first real darkness, though, they slip below, where at once they turn into the tiny phosphorescent granules foreigners mistake for dinoflagellates. Gradually they sink, and are re-born into fry of their original species.)

From afar the fisherman had been sitting reflectively in a dreamy cloudscape. Now that we were close, the angle of vision had changed and the pinkening city with its ever-excluding portals had receded, leaving him starkly present and casting his own shadow. He had acquired power. His first act had been to summon us to him from over the horizon, his second to have thrown us into confusion. Now he presented us with a dilemma. The things which people said (and which you will certainly be able to recall since you yourself said some of them) included:

'We don't know who he is.'

'We haven't the time.'

'If we touch him we'll have to get rid of him. *And* the boat. They'll pin it on us.'

'How do you know he was murdered?'

'We can't just leave him.'

'I say leave him.'

'He must have a Christian burial.'

'Are you a priest, then?'

'How do you know he isn't a Moslem?'

'Ho, are you an imam?'

'We really haven't the time. The sun's setting.'

'At least let's find out who he is. We can let his family know.'

'Volunteering, are you? Perhaps he died of AIDS. You want to touch him?'

'We can't just leave him.'

'We haven't the time.'

'It'll bring us bad luck if we don't.'

'It'll bring us worse if we do.'

'I say turn back.'

'We haven't the time.'

And meanwhile we kept describing slow circles in the water around this monument, following our own fudged wake in a circularity of thought.

(Other spirits of the grilling fish do not immediately return to the ocean, but continue to hover about the fisher-folk. They are not looking for a chance of revenge. They are held to be the Earth's original inhabitants, and, as such, are too ancient and wise to indulge in human passions. Instead, they look for sleeping children under the age of five. Unseen, they slip between the parted lips and take up residence in their little hosts. Such children can always be told apart, for they are constantly in the water and cannot live anywhere but by the sea. They will become fishers themselves; they have no choice. And sooner or later, by one means or another but usually by drowning, the fish/man-spirit returns below the waves. Thus an endless cycle is joined, of man and fish alternately pulling each other up into the air and down into the deeps, locked into a sublimely inscrutable continuity. No one claims this process does not involve suffering.)

Even in these regions the dead are encountered with a frisson. People under the influence of startlement usually say nothing unexpected; in terms of what is expressed, death merely jars loose the predictable; this alone being the evidence of something else held in abeyance. Nobody observed of our fisherman that he was the colour of clouds. Hypostasis had caused his blood to sink so that the upper half of his body was wax-pale. Still less did they say his spirit had powerfully engaged with us and, no matter what we did now, was setting the terms of our entire trip and maybe even those of future ones.

There is an alternative theory, of course. There always is. Justus Forfex (Giusto Forbici), quoting Alzina (1688), tells of a belief '. . . *que el alma nacía y moría nueve veces . . .*' The soul becomes progressively smaller with each cycle and after the ninth incarnation could fit a *longon* or coffin the size of a rice grain. The body, on the other hand, changes into water at night but by day lives with the soul. This lone fisherman's soul was patently unshrunken, monstrous in its presence, so he must have been in an early stage of his nine lives; while his unburied body equally clearly yearned for the water which oncoming night was already darkening about and beneath us. Meanwhile, our fear and uncertainty made us dither and go on drawing and redrawing the tedious outline of the problem he had set us.

During this debate the captain stopped the engine to save fuel. The ring of silence which surrounded the dead man expanded. The dark triangle of his hat pointed at the city still expanding in the sky behind him in recklessly unplanned development. By now it was a fantastic conurbation full of immense gargoyles and gothic high-rise blocks, all stately plasticity and picked out in salmon and silver. Its vistas of serene collisions offered the same haunting promises of ways into its interior through a labyrinth of crevasses towards an illusory heart. As with a starfield, all patterns and perspectives were relative to the viewer's position. Seen from the other side of the horizon it would have been a different city or not even a city at all, nothing but lumps of vapour and scattered bags of dew. In any case it was a city built entirely of water. Its towers were water, its colonnades were water; its arches, lintels and architraves were slowly tumbling water. Its foundations were planted in more water, a polished ocean still placidly reflecting the liquid mass of its construction. This dual image now met itself in scarlet at the horizon in a way that suggested the dissolution of one into the other in a huge collapse of red rain somewhere far over the planet's shoulder.

At one point the *Bhavisyottarapurana* exclaims 'Water! Thou art the source of all things and of all existence!' This intuition probably influenced Thales of Miletus when he became the first Western philosopher to attempt a rationalistic account of the phenomenal world. He took the received idea of there being four elements which composed all matter and reduced them to one by claiming to show that earth, air and even fire derived ultimately from water. Being a philosopher, Thales was keenly aware that not every liquid *was* water, for it is reported that by buying up olive presses the year before a bumper crop he made a killing in oil. This would have been in about 580 bc. Approximately 2,565 years later, the captain of our boat resignedly tied a T-shirt over his mouth and nose and lowered himself from the prow into the dead fisherman's craft, which skidded lightly away with him to a distance of several metres. For a second his masked head passed across a slum district of the aerial city.

We watched him gingerly approach the sitting figure from behind. Maybe his stealth was from having to keep his balance in a tiny craft designed only for a single occupant, his high centre of gravity making the bamboo outriggers alternately rise and smack. From where we sat, his hesitant progress looked as though he feared the fisherman might suddenly turn at the touch of a hand on his shoulder. In the silence we could hear our captain's low mutter behind the mask: Catholic prayers or else respectful requests for permission from the sea spirits clustering around the little boat. Perhaps he was simply talking to the fisherman into whose face he was peering, apologising and asking forgiveness for crouching down to put his hands briefly into the pockets of his shorts. When the fisherman had set out on his last voyage the shorts would have been loose. Now his distended stomach made it hard to perform this intimate search. In a moment the captain was rinsing his hands over the side and standing up again.

'Nothing,' he called. 'Not even a cigarette.'

'No blood?'

'Not a mark on him that I can see. There's nothing in this boat except for him and this fishing line' – he lifted a short length of thick bamboo around which the nylon filament was wound. 'Oh, and a baler.' Delicately, he stretched past the man and retrieved a red plastic pot, the ubiquitous 1-litre engine oil container with its top cut off. Then he hesitated before again reaching between the sprawled legs and straightening up with a piece of paper in his hand. He put this in the pot and made his way back to the stern, where he lay and began paddling with both arms. So light was the skiff and so calm the sea that he and his dead passenger were soon alongside. Our captain reached up and swung himself on board the *Medevina*. At his involuntary kick the fisherman once more scooted away, adrift and rocking.

The scrap of paper was blank. It had been torn from a school exercise book, lined in alternate pink and blue feint.

Creases suggested it had been screwed around a small contents—pills perhaps.

'Heart medicine,' somebody said.

But you will remember a voice at once said, 'No,' and faces turned at the speaker's apparent certainty. 'No, because there's no bait, is there?,' which made the captain glance sharply towards the drifting back beneath the straw hat. 'Who goes fishing without bait? What does a bare hook catch?'

'Perhaps he kept it in the baler and had finished it. In any case he's fishing, isn't he, Mots? He's holding the line.'

Mots was stubborn. 'Then where's his catch? He's not fishing. That's not what he's doing.'

'He's not *doing* anything.'

'Yes, he is. He's being dead, just as he intended. The paper contained poison. He came out here to die, only that. So naturally he brought nothing else with him. After a while he must have thrown the paddle away and just drifted. Boat empty, pockets empty. Why catch fish you'll never eat?'

All the while the sun was sinking amid the city's vermilion and cinnamon rubble. Soundlessly the towers and palaces had been shaking themselves down, flattening into featureless broad terraces of farewell colour through which the sun settled like a marble in oil. Behind us night's tardy priest was hurrying up over the bent Pacific. In silence we watched the dead man darken until he became an outline, an emblem, a timeless and purposeful shape, the patient unmoving fisherman. Without warning, without word or grunt, our captain started the engine and in shocking blare we resumed our course.

There was no one aboard whose gaze was not fixed aft. For a great while, it seemed, our wake went on opening up between us without the fisherman's figure diminishing. Then suddenly he had shrunk from a monument to a miniature chessman, whittled away as much by the withdrawing of light as by increasing distance. There came a brief period when he would vanish and reappear as the invisible swell raised him high enough to catch a last smoulder. Then he descended into troughs and wells already pooled with night. Even when he dissolved entirely and was lost, our eyes still watched the place like Sir Bedivere straining for a last glimpse of his dead king's boat.

'I still think we should have buried him,' a voice said at last above the engine's din.

'I don't think he committed suicide at all,' someone else said indignantly. 'He was killed and robbed, like always. There're lots of ways of killing a man without blood. Pirate scum. They stole his catch and everything else, but the boat wasn't worth a tow so they just left him.'

'But *we* oughtn't to have left him . . .'

'We didn't have the time. We're already late as it is.'

'This trip was jinxed from the start.'

'We had no time.'

'No time.'

III

They say death has changed for modern man, that it has been deconstructed and, like him, become postmodern. Sometimes, when the day is bright and blue and hot enough to be quite empty, and the rocks shimmer in the sun, the conviction comes that at their heart human societies are just elaborate fabrications for suppressing a knowledge of death—conspiracies sufficiently complex and beguiling that the dark secret of our own mortality no longer obtrudes. This huge artifice protects the race against its Achilles heel: the certainty that all its affairs are nothing and soon it will be one with the dinosaurs. It is as though it had decided to club together and invent a series of wonderful tales, from Gilgamesh to cyborgs. The principle is that of the Arabian Nights, with everyone their own Scheherezade.

On the morning wind we seem to hear the creak of a million treadmills, the squeak of rowing machines, the trilling and drilling of an endless aerobics class. It is the dawn chorus of anxiety. Another kind of insurance is being enacted, that private/public investment in keeping fit and being seen to be keeping fit. Apart from exacting its own toll in humourless tedium, it turns ill health into a personal failure such that death is seen as just deserts for not having taken the trouble to be sufficiently alive. The body as machine, the unread user's manual, the culpable lack of maintenance: they all form a nexus of irresponsibility and downfall. Someone fails to turn up at the gym as usual in their Lycra leotard. After a few days their name

escapes us. It is understood there was always something more
they might have done: another few metres' jogging a day, many
fewer beers and cigarettes, a further notch of health reached in
order to carry on being fit indefinitely. (What was it we failed
to grasp even as we hung punitively from wallbars? Can the
mind be rotting atop its splendid torso?)

In what pathetic fragments we move, believing ourselves
whole! The precious 'I' disappears for long stretches of each day
and entirely vanishes during sleep. In one of our registers something never forgets that in default of the ocean deeps, a refriger-
ator door is always yawning for us as the prelude to spade or
flames. Here, at least, the old mythologies no longer work as
they did. It is not possible to envisage a private survival. We cannot believe as Sir Thomas Browne did, the same scriptures lying
open before us with their panoply of promises. There can be no
magic left in prophecies of paradise. That all life was held to
have begun in a garden and—if we are good—will likewise end
in one convinces nobody. The majority of the world's people are
now town-dwellers to whom rural metaphors are no longer
instinctive. Since most people who imagine life after death think
of their ordinary lives transfigured, a townee would find the
myth of the celestial city more plausible than that of a garden
paradise. And what might an urban paradise be? Strange that
Scheherazade's own name meant 'city-dweller' in Persian.

There is pathos in the way religions of the book have
become immovably beached on the littorals of faraway and long
ago. To desert-dwellers, what more natural than to see heaven
as a sublime oasis which owes its existence to nothing more
mystical than *water*? Here we are, the deserving, eternally at
peace in a lush garden, sprawled in the shade and recovering
after life's gruelling journey. 'But the true servants of God shall
be well provided for, feasting on fruit, and honoured in the gardens of delight. Reclining face to face upon soft couches, they
shall be served with a goblet filled at a gushing fountain, white,

and delicious to those who drink it. It will neither dull their senses nor befuddle them. They shall sit with bashful, dark-eyed virgins, as chaste as the sheltered eggs of ostriches'.[1] Ignoring the politics, we note only the continuing presence of physical appetites, the slaking of bodily desires. The twelfth-century philosopher and physician, Moses Maimonides, was impatient with this sort of naivety. 'To believe so is to be a schoolboy who expects nuts and sweetmeats as a reward for his studies. Celestial pleasures can be neither measured nor understood by a mortal being, any more than a blind man can distinguish colours or the deaf appreciate music.'

The essence of paradise will always be conflated with that of lost Eden since the future is unimaginable and the present unmythic. The very word 'paradise' comes from Persian via Greek and means a park. What could the modern world offer by way of a matching tranquil and timeless vision? Are we to recline for ever in some leafy municipal square, where the sun filtering through the trees dapples us in a bearable radiance, where traffic noise has ceased, where litter-free paths are strolled by the righteous eating ambrosial hamburgers? It doesn't work. Nor does a southern Californian dream of bronzing our cancer-proof skin beside Hockney-blue pools, endlessly dating and mating and clinching deals. Besides lacking depth, such visions have no ecstasy. In any case, our bodily needs are now largely catered for. It is impossible to imagine any central image as simple and important as *water* which might resonate for us as a condition of life itself. And this is the triumph of material mastery, that it supersedes and blots out the symbolic to the extent that the only resounding things left are absences. Nature has fallen beneath *Homo*'s power, and in doing so has left him without an image of heaven.

[1] Koran, 37:48.

In distant archipelagos there are often days of heat and dazzle powerful enough to wipe out thought, to leach away everything that is not planetary furniture: trees, rocks, clouds and water to the horizon. To people who skitter about these gulfs of ocean in pea-pod craft—a few sheets of marine plywood tacked together with copper nails—death is an imminent presence. So it is for those hacking at their stony fields high above the glittering shoreline. Subsistence throws things into bright relief. At the end of the day's work either there are fish and maize cobs to be laid over charcoal or there are not. Infants are born, linger for a week, vanish into the ground. There are no hidden deals. Everything—shelter, food, water—is in the open. The facts are dealt with by private treaty. The woman pounding her washing on a stone in the stream, the plodder behind his buffalo knee-deep in a rice paddy, the lone fisherman far out in the sun's glare: each might harbour wistful dreams but none pretends to be proof against the cancellation which can descend in an instant, whimsically, without notice or reprieve. The beaches on which their children romp and tumble are composed of the dead. The whole landscape is a cemetery. Diatom, mollusc, foraminifer, cuttlebone, dog jaw, pig tooth, and all by the trillion ton. Daily they walk on the deepening past, mend nets on it, fall asleep on it, make cement blocks with it, shit into it. The sea turns over and over, a geological machine smoothly meshing its gears and grinding up time itself. At night it sparkles with energy. Sit beside it under the stars and fan the driftwood embers and watch very seriously the broiling fish as slowly they curve upwards from the heat. Not as dead as they look, some of them, for now and then one leaps off the coals the instant it is laid across them. '*Buhay pa!*' and a child's delighted scurry to retrieve it and place it on the fire again. 'Lie down! Go to sleep!' Companionable laughter. The incense of smouldering fish oil drifts across the constellations and brings the village dogs from far down the beach.

Despite the skinniness of this living, for all its rigour, there can exist in these rural and marine backwaters a certain ebullience. Not to sentimentalise it, it is that peculiar freedom which descends like a gift on those so constantly menaced that they slip off the burden of mere worry. A strange security results when death is so close a companion. It can be felt while crunching along a beach, a skeleton walking on skeletons with the time machine turning in step, wave-falls and footfalls. A gleeful levity at being so brief, at feeling so exempt. Freedom from what, then? Certainly not from the irreducible pact of living. Rather, from the heaviness of having to share in metropolitan anxieties, the contaminating conspiracy, the yearning. Freedom, too, from the corroding suspicion that the extra time bought by wallbars has already gone on wallbars. The Thousand and One Nights (*alf layla wa layla*) never meant a literal two-and-three-quarter years but was just an expression for a long time, the eternity spent in staving off the executioner by babbling at him.

IV

On into night we went, nakedly starlit, without even the stealth of a light overcast. The equatorial sky blazed with local stories. Little Turtle lay helplessly and wept a stream of stars which fell beyond the sea's edge. She had been overturned by the turbulent rush of Tig-it, the giant manta whose eye—to other folk in distant lands—was Achemar in the constellation of Eridanus. Little Turtle wept because, being on her back, she could now see only the darkness of Earth and nothing of the guiding sky with its brilliant lamps and beacons and knew she would never reach home. Far away from her was the Brazier with its frozen rag of ionised gases drifting for hundreds of parsecs, through whose smoke Cry-Baby was being passed three times to dry his tears. Into the Brazier had been sprinkled incense, seaweed, fish bones and chips of scented wood. The enchanted perfume filled the heavens, reaching the nostrils even of the Octopus. The Octopus was sunk deep over the horizon, biding his time. He waved his tentacles menacingly, their suckers marked by stars. Now and then, sparks from the Brazier shot across the sky. Whenever these meteors fell into the sea they changed instantly into pearls which sank, still glowing, to await a lucky fisherman.

Over everything such orderly dramas shed a stark, revealing glare, a million suns lighting up a single Earth night in which a fishing boat was hoping to poach some lobsters from another nation's territorial waters. The noise of its engine reached all parts of the universe, like the Brazier's incense. You will remember that sensation, something of stateliness as well as apprehension. The fear of coastguard cutters, naturally: of an impounded boat, confiscated tackle, jail, diplomacy, bureaucrats and fines—whatever else, the fines. But behind all that, like the Octopus hunkered down in his lair beneath the horizon, lay the shadow of the man we had left in his boat. He was there somewhere behind us, still sitting in his broad-brimmed hat, the fishing line with its unbaited hook trailing from his dead hand. Unwaked and unburied, far out beneath the stars he sat, lifting and falling in the long Pacific swell. So must he have sat many thousand times before. Nights of moon, nights of rain, nights of angry chop sending water spilling over the low freeboard, obliging him to loop the line over a copper nail and set to bailing with the plastic motor-oil container. Nights when he could see into the depths with his eyes shut, sense the eddies of 10,000 tails and drop his hook right among them. Cold, too, on other nights, sitting there in the empty wind, now and then fumbling a cigarette alight with the tiny companionable flame which flickered deep in a mayonnaise jar.

Some nights he and his colleagues would remain within hailing distance, their yellow lights so low in the water they winked out in the troughs of passing wavelets, then on and again off, with the steady persistence of fireflies. The water beneath them might have been empty of life. As if drawn together by a mutual acceptance that no one was going to get fat on this night's fishing, the boats stayed within conversational distance. Gossip was exchanged, cigarettes, matches, a spare jersey. A flat bottle of rum was passed around. There they would sit in that enclosed world which sprang into being, even in the nowhere of mid-ocean, when men talked like survivors of a series of grim accidents ashore. The sicknesses, the deaths, the separations; the plot of unyielding land whose scant crops were tithed by a landlord; the truant son, the feud with the policeman's family. Over and over again stories were told that everyone knew. Their retelling, and the long reflective silences falling between each, were a description of that other world whose reach seemed absolute, yet outside which it was still possible to sit. The lamps of land could not always be seen. The fishermen's night-lights danced their flames within their sooty, improvised chimneys, marking out the coordinates of an existential annexe as temporary as it was shifting. Within it, something invisible and unspoken came into being. For a while the flames were the bars of its cage and it could not escape until the cage fell apart or the blanch of dawn disclosed a loose knot of tired, stiff men floating on the sea. This unspoken, this invisible—which sometimes smelled of tobacco and sometimes as though incense and fish bones were burning—was the pact. It was the silent deal done with the sea and the stars, with what welled up from below and sifted down from above, with immeasurable gulfs. The deal could never be expressed. Into its secret balance nothing weighty was thrown, simply the piecemeal fragments of what each inner mind held in safe-keeping: a particular smile with a hand half raised in neither greeting nor farewell; a receding wave draining sand and frag-ments of shell from around bare feet and undermining the toes crawlingly; the shimmer of a cooking fire at dusk and a child's laugh; the gecko's grating call—the cry the dinosaurs knew and which would not stop until after the human race, when the planet ran down: '. . . loves me, loves me not; loves me . . .' (as European children once blew dandelion clocks). And heaped into the balance as these fragments might be, the deal remained always the same, imperturbably. The Octopus's suckers twinkle; his arms await. They embrace the sky. The playlets overhead are scanned for portents, warnings and news of all kinds. The playlets enacted below are watched by no one and nothing.

All these things were turned over by the living about the dead, until the lone fisherman bobbing far behind at our wake's end grew in stature like a fable. There was not a soul aboard without guilt for the uncompromising corpse we had left. His boat was clustered around with spirits: they had been practically visible, perched like transparent vultures along the outriggers and fidgeting at bow and stern. Some would have jumped ship—would even now be swaying atop the black poles supporting *Medevina*'s tarpaulin roof or hunched on the fuel tank of her temperamental Briggs & Stratton sixteen horse-power motor which had already caused us delay and, causing that delay, had engineered the encounter with the unknown suicide who had now adopted us. What trip starting thus could turn out well? So you will no doubt remember the thrill of portent with which you abruptly discovered that the bank of cloud low above the sea to starboard was land, that we were now in foreign waters, that a new playlet was about to begin.

When the motor falls abruptly silent it is the time for stretching, for scanning the scene with ears as well as eyes. No cutter's masthead light crawls across the nearby bulk of cliff, no angry hum of motorised pursuit. Just the slap of water under prow and outrigger noses. Our intermittent, eventful journey has brought us here, late; our stubborn captain's calculations can almost be heard. Half-past one: tide change in thirty-five minutes. We'll need the anchor. Throughout these archipelagos, in a thousand straits and sounds, channels and passages, the current switches direction. Twice a day it streams, races and eddies, often forming boiling demarcations where two contrary currents intersect. To sit there with divers deployed in relays, working the lobster beds, the boat must remain on station. It cannot be allowed to drift. Our captain must have Bajau blood, they say. He does not even possess a compass, yet he has brought us back over his secret patch of seabed after almost twenty hours of travel. Nobody aboard has the slightest doubt that when the first two men go down, their torch beams will fall on familiar terrain.

The anchor is thrown over from the prow. It is made of inch-thick reinforcing rods stolen from a construction site, knobby with crude welding and rust. Two men paddle in the slack water to bring the line taut so that when the current begins at tide-turn we will already be nosing into it and can drift no further. With a tyre lever a boy is slipping off the frayed fan belt which drives the propeller shaft. He eases it instead over the pulley of a compressor bolted to the deck-boards. This, too, is a rusty lump of engineering. It is connected to a converted cooking-gas cylinder with a twin-nozzled tap. Men are breaking out two coils of thin polythene tubing whose insides are mottled with mould, pushing the ends over the tank's outlets, tightening clips. The first two divers are already in the water. Their wet knuckles gleam as they hold on to the boat. In addition to T-shirts and shorts, they are wearing home-made goggles carved from wood, the eyepieces set with a lozenge of glass embedded in pink marine epoxy. They carry torches waterproofed by a motorcycle inner tube and lengths of stout wire with one end bent into a sharpened hook. A lidded basket on a rope is handed to each. At the captain's signal, the still-hot engine is swung, fires, backfires, splutters, blares into life. On its continued running the divers' lives depend. They will be working at between sixty and one hundred feet. Lobsters are most abundant at about 1,000 feet. Puny half-naked men with ramshackle compressors and cheap Chinese flashlights can do no more than skim off the topmost handful, concentrating on the two main species which can be found on these comparatively shallow reefs: slipper lobsters and spiny lobsters.

The Briggs & Stratton has been throttled back to a slow blat-ting. The air hoses are tossed to the men in the water and thrash about, fizzing, until caught. Each man folds over the last inch and grips it between his teeth, regulating the air pressure by clenching or relaxing his jaws. In the water's uppermost ten or fifteen feet they will have to bite quite hard; below that, their grip must gradually slacken to allow more of the foul-tasting air to gush into their mouths and counteract the sea's growing squeeze. Looking over the edge of the boat, the rest of us follow the small, wavering green puddles of their torchlight until they are extinguished in the blackness beneath.

We who are left behind. The phrase carries both anxiety and mourning, ready to shift emphasis in an instant, and into the bargain flies a modest pennant of heroism. The sea captain's wife, her dark-lantern casting up her strained features, paces the widow's walk on her roof and stares into the gale for her man's returning gleam. The ground crew watch in silence as the departing squadron lifts from a darkened airfield somewhere in wartime Europe with its cargo of bombs and boys— teenagers, some of them, children sitting or lying prone in vibrating alloy nests. And we, too, go on staring at the patch of sea beneath which our friends have vanished, each mind anxiously drawing on the black surface its own sketches of misadventure. The direction of the tubes and ropes gives no real clue to where the divers might be. The two older men paying out the coils have already put overboard fifty metres apiece, passing it steadily through their hands and easing out kinks as it goes. They are waiting for the tide change, to see which way the boat will swing. The divers went over on the starboard side, the tubes vanish over the stern. Suddenly the men are paying out to port, and, with a shock of re-orientation, you discover the dark mass of land now lies off to the left.

Can you remember how many times you have watched this? And have yourself gone down with that hissing pipe clenched in your molars until your jaw aches? The torch beam hollows around itself the pale black cell in which you descend, heading down and down. Twinkling dinoflagellates bounce off your bare hands and arms in crumbs of cold light. The spirits of martyred fish, they stream from the invisible plastic tube as though on your trajectory through space you had crossed the path of a meteor shower. Head-down through these summer Perseids you descend, eyes behind the oval glass peering for focal points, imposing patterns on nothingness, until one of the patterns catches. It gives you a queer pick up at the heart to see it harden into a detail of rock with small nocturnal fish going their ways, a cantle of this universe which hides itself from the world above. This place is not simultaneous with the realm of air. It can never be reached by anchors and plumb lines. It happens at different wavelengths and is full of a vivid-ness that remains invisible to the air-breathing human, his ears plugged with water and his eyes as dim as his weakening torch. He is an interloper in a land full of playlets he cannot quite see and stories he just fails to hear. Palps and antennae winnow pheromones from the passing currents. Unsensed, a teeming sexuality surrounds this visitor, who can smell nothing, nor taste anything but mould and exhaust fumes. All he can ever bring back from this ancient domain are beautiful banalities like pearls and lobsters, and even these are more beautiful than he knows.

You are not going down tonight, though, but instead will sit in the darkened boat with the others left behind. After twenty minutes, the first diver emerges. His basket is hauled aboard and emptied in a cascade of claws and carapaces.

'Okay?,' our captain calls to the man in the water, who has removed his air hose and is taking breaths of clean night air.

'We're behind the Madonna. The first of those gullies. It's good.'

'One more and then Mots goes down.'

The diver's pale hand waves, replaces the air hose, grabs the empty basket and drags it under the water. A single plywood flipper thrashes the surface briefly and he is gone in a swirl of luminescence. The boys begin picking lobsters out of the bilges, tying their claws with a double loop of plastic straw, packing them into the first of the ice boxes. Among them is the flaccid mass of an octopus. By starlight the boys look for the marks left by the diver's teeth where he killed it. They are difficult creatures to tease out of a hole and nobody would waste time doing so on a lobster hunt. The diver's light must have caught it in the open, a dark pink and pallid umbrella flowing desperately over the exposed coral, tentacles searching for a crevice into which to furl itself. It is possible to take them then, to grasp the bulging pulpy head in one hand and bite the narrow neck which joins it to the tentacles. The flesh is tough and rubbery and our diver would have had to remove his air hose before chewing at the creature. If you can detach the octopus entirely from the rock before killing it, it will clamp its arms about yours, pulling its hooked beak towards your hand. Biting at this fleshy club is even less pleasant since it immediately transfers some of its sticky grapples to your head, welding itself over your face as you try to get a mouthful of the vulnerable part. Yet an experienced fisherman can kill a small octopus with his teeth quicker than he can with a knife, for even the sharpest blade can skid off the gristle and gash his own hand.

The second diver soon surfaces and his basket is emptied. 'Not far from the Madonna,' he says as he disappears again. The 'Madonna,' we all know, is the name someone on a previous trip has given to a curious coral formation, a slender pillar some fifteen feet tall which in outline, seen from one aspect, resembles a statue even to the extent of having a texture like the petrified folds of a robe. On her seaward side, the reef falls away into a series of steep gullies, in the first of which our divers are working, and she forms a useful marker.

When each man has filled two baskets, the divers come in and are spelled by a second pair. Our captain is strict about this. Each pair dives for roughly forty minutes, rests for the same, then goes down for a final forty minutes. Whatever theory he is working on, it bears little relation to the rigid decompression rates on which an alien medical science insists. Transient headaches are a matter of course, yet nobody in his crews ever gets the bends, though there are crippled young men further up the coast. It is mysterious.

The engine chugs on, the compressor's worn cylinder clatters, the valve on the tank bleeds with a chattering hiss. The boys go on packing the scrabbling lobsters in ice. A loud sizzle is heard aft. In the galley (a sheet of tin on which stands a cement cooking stove large enough for a single pot) someone has tossed the first pieces of chopped octopus into a work of hot oil. The domestic smell of frying drifts across the boat and over the sea. Faces are dimly illumined by the red glow of the stove's mouth, across which a hand with a palm-leaf fan flaps rhythmically. The captain and others go on scanning the darkness for the first signs of a coastguard cutter. Yet the deployed divers and the cookery would make a quick getaway impossible, and, after the initial tension, a calm insouciance overtakes us until we almost forget the dead man over the horizon.

V

When he quoted the Jesuit, Alzina, Justus Forfex was beginning his short study of sea burial. As he was to relate in his autobiography *Viaticum* ('Life has written me, not I my life'), he was shipwrecked eight times in his wanderings between 1842 and 1867. One way and another, he was able to observe burial customs from Subang Gulf through New Castile to what he whimsically dubbed 'Cholynesia, or the Bile Islands.' He also had the opportunity, over long months' tattered living off crabs and coconut juice on deserted atolls, to reflect on how the significance of water pervades both soul and body.

Years later, as Giusto Forbici, he passed his scholarly retirement in a gaunt, book-lined palace in Arezzo. When asked why he had chosen that particular city (he was born on the Italian coast, down south in Salerno), he replied that there was no part of Arezzo from which one could possibly see the sea. In his most notorious essay, 'De atramento et oceanis morteque,' he writes of the relation between ink, oceans and death. For those with a little Latin, his title makes the connections less apparently strained since the word for ink, *atramentum*, would call to mind the adjective *atratus*, meaning 'darkened' or 'wearing mourning.' This essay had its roots in the circumstances of Forbici's last and most desperate shipwreck, when he was the sole survivor washed up on a coral strand in the heart of 'Cholynesia.' Washed up with him was more than his fair share of irony: nine immense straw-covered glass carboys which he remembered having glimpsed in the ship's hold. These, he knew, contained fresh water. His elation at this reprieve was short-lived. They turned out to be full of ink, part of a consignment destined for Dutch bureaucrats in Batavia. The ink was made of dried sepia extracted from thousands of cuttlefish, purified, mixed with lamp-black and thinned with spirits and water. It smelt of bad fish. Forbici thus found himself alone on an islet with many gallons of a substance scarcely more drinkable than the ocean surrounding him.

With immense effort he trundled the great flasks up above the high tide mark and buried them. He hoped to prevent the direct heat of the sun from evaporating the alcohol through the cork bungs and thereby stop the liquid's fishy component from putrefying still further. For nearly three months he eked out survival by drinking small tots of rotten ink. They made him retch as well as leaving a slight hangover. He produced occasional sooty turds full of free carbon, until towards the end as he grew weak and feverish he was seized with stomach cramps accompanied by a stygian leakage. Not only was he drinking ink but seemed to be excreting it as well.

How formative is illness for the imagination! Entering what should have been his final delirium (he later wrote), he saw himself as quite conventionally encompassed on all sides by death. Yet suddenly he perceived it was a distillate of that very death—as produced by octopuses and cuttlefish—that was keeping him alive by a kind of homoeopathy. Was this proof of the eccentric Dr. Hahnemann's 'Law of Similia'? How did these cephalopods extract their black essence from clear seawater? Since it was impossible to make an element such as carbon, did this not imply that the world's oceans already consisted of highly dilute ink which the creatures then concentrated? Two-thirds of the planet was a fatal inkwell.

These thoughts seemed to him very clear and satisfactory as he sat under a sparse canopy of thorny macrophyllum whose mildly hallucinogenic leaves he now and then chewed. Much later, after being rescued and having recovered his wits, he remembered Varelius's little monograph on the presence of gold in the sea. Forbici then made the leap which was to give his own essay its fame. He reasoned that if the sea acted as a body of solvent, leaching out of rocks their auriferous veins, then its clear depths must contain all the essence of land (*il nocciolo della terra*), down to its last molecular constituent. If the human body itself was composed only and wholly of chemicals and minerals that occurred in the Earth's crust, then it followed that the sea must also contain all the component parts of a person. In theory, therefore, a wizard, alchemist or divine agent would be able to assemble a human being from a quantity of seawater. Nothing would be lacking but the soul; and who knew but that also might be found lurking in the invisible interstices of water?

This lively fancy was not at all understood in Arezzo and went down positively badly in Pisa, Siena and Florence. It was scarcely ten years since Darwin had published his own genealogy of the human race, and that had been impious enough in its proposal of a scurrilous ancestry that vastly pre-dated the book of Genesis. Now here came this vagabond—an Italian no less—who was just as godless but even bleaker, to suggest that the frame made in God's image was composed of nothing but chemicals after all, regardless of its lineage. The Bishop of Lucca referred in a sermon to 'a fashionable, Satanic materialism that . . . reduces the sacrament of motherhood to the level of a factory or a workshop assembling puppets,' a remark which some believe gave Carlo Collodi the germ of his fable 'Pinocchio,' which was published a decade later in 1883.

Forbici, leathery and reclusive in his Aretine library, gave not a fig for the Bishop of Lucca. 'I look in the mirror and see—nothing,' he wrote exuberantly. 'I look through the casement at the people who, from the sound of their voices, I judge are in the courtyard below but who seem today not as visible as they should be. Our bodies—what are they? Diffuse clouds of whirling atoms held together by some clotting or electric propensity. Why should this appear so miraculous? How can we ever embrace? It is our brilliant *fluidity* that provokes me to love whatever our poor shadows extend. We pontificate and preen, we survive childbirth and shipwreck with all the corpo-reality [*fisicità*] of a cloud. The sky-cities we saw above Pacific horizons have long since returned to the ocean, and so shall we. What is it about this process that makes one confident and affectionate? Cradles full of clouds. Coffins full of clouds. Shaving-mirrors full of clouds. Books full of clouds.' This retired traveller then went on to meditate—at too great a length, maybe—on water. He returned to the relationship between dilution and concentration. Was it possible for *un nocciolo* to be dilute and still remain an essence? Can a con-centration be said to exist even when dispersed among extra-neous things? It is hard for today's readers to address such questions with Forbici's tirelessness. His staunchest admirers are obliged to admit that in his later, landlocked years, the great Justus Forfex became obsessive about water, bodies, the oceans, ink and so forth, returning to them in essay after essay. He never did manage to pull them (or himself) together into a single philosophical proposition. Instead, they flow in and out of his thoughts and writings like busy animalcula crossing a microscope slide. Cholynesia had taken its toll.

His earlier monograph on sea burial, though, showed him at his best. It is a lively volume full of arcana and jolting details, still widely consulted by anthropologists, scholars and ghouls. It lists with great attention to detail the classical or ancient practices as well as those he had personally witnessed in the southern archipelagos. And yet even this essay lacks a central thought as to what it might mean to be returned to water, as opposed to the magic benefits accruing to survivors for burying a body in the sea. 'Returned to water' is the significant phrase, one which might not have occurred to him then because this little book antedated his final and most traumatic shipwreck. He had not yet formulated his thesis that all bodies are by nature oceanic and that earth-burial, urn burial, cremation, etc., are merely forms of postponement. (Even fire is water.) And, of course, he was unable to end his monograph with the highly pertinent story of how he had finally been rescued from his atoll with its dwindling ink supply.

Forbici dealt with all this in his so-called 'Parergo,' a 'by-work' or corollary to the rest of his writings. When Marcello Vanni came to produce the collected edition in 1907, he judiciously printed the 'Parergo' twice, inserting it as an appendix to *Viaticum* (where it forms a neat climax to his author's adventures) and again as an addendum to the monograph on sea burial, *Talassotafia*. In this position it augments the list with one last strange example, while the autobiographical element gives it an immediacy which is irresistible.

Over the long weeks of isolation on the atoll he had undoubtedly entered an extreme world. There were times when all his shipwrecks and misfortunes collapsed together and he knew he had been living on the islet for centuries, slowly wading its leeward shallows in a constant quest for food. He had reached the stage where the finding of a single overlooked sea urchin became the high point of the day. Using probes of antler coral he would prise open the friable shell, trembling, so as to leave the inside undamaged. Then, having reluctantly discarded the animal's rudimentary digestive tract (which only made him vomit ink), he greedily ran his thumbnail around the shell's interior, scraping up a tiny mound of gonads which tasted to him like the rarest caviar. The ghost crabs which lived in burrows in the coral sand and danced in and out of the rippling wavelets at dusk were still plentiful but difficult to catch. When by guile (blocking up its bolt-holes) he did succeed in catching one, he triumphantly ate it whole, crunching up claws and carapace even as the translucent running legs scrabbled at his lower lip.

He had tried and failed to catch fish, even the tiniest fry that hung like flakes of brilliant metal about the coral outcrops. No net, no line, no hook. The scholar was reduced to dropping boulders into the shallows where the fish flickered in the hopes of stunning one. The turbulence subsided, the water cleared; nothing had changed except the new naked rock lying there with the braver fish already nosing around it. Forbici did find one place on the atoll that looked promising, a pool among the rocks on the windward side. Roughly circular and about four yards across, its ragged sides fell vertically for several fathoms, forming a kind of porous shaft. Over many centuries, some freak of the current must have scoured it out of the dead coral formations of which the islet was built. By lying at its lip and putting his face close to the water's surface, he could see, beyond his own ragged castaway's visage, a dark bottom strewn with broken coral: white chunks and lengths among which bulky parrot-fish moved lazily.

For long hours he pondered a way of trapping these succulent creatures. With some damage to his hands he managed to break off several wands of the tough little bush which had become his shelter. These he tried to weave into a trap of sorts; but no sooner had he dipped this crude, lopsided construction into the water than it began to unravel. Besides, as he realised too late in an outburst of rage, he had no cord with which to lower it, and nothing tempting with which to bait it. He gave up and went back to eating titbits of weed and mollusc he found beneath the rocks in the lagoon.

Otherwise the days passed unrecordably, and the listless view from beneath the macrophyllum repeated itself at the mesmeric pace of evolution itself. It is not difficult, even well over a century later, to be confident of what Forbici saw. On the atoll's windward side the craggy foreground hides the water's edge. Impelled by the long Pacific swell, occasional tufts of sea leap up among the coral rocks into the dazzled air. A marine heartbeat thuds underfoot and trapped air sighs gustily in vents and fissures. The water in the pool rises and then sinks. This is land no solider than a petrified sponge, with the sea passing continuously through its roots. Small crustacea move on the lunar surface; the isopod *Ligea* scurries everywhere into cracks like a littoral woodlouse. It is a fossil scene.

At night, according to Forbici, a cool breeze sprang out of nowhere. In an act of self-burial which might bring him a little warmth he would heap up the detritus of a billion dead creatures over his legs, for he was sick and shivery. He had nothing to do but lie on his back and stare at the universe. Forbici's cosmos was entirely classical: Greek and Roman and Arab. Apart from his training as a scholar, years of voyaging had left him with a good knowledge of the heavens. The pictures he saw were familiar: an archer, a herdsman, Hercules, a plough and a lyre and Berenice's Hair. To stop himself falling forever upwards into the radiant silverpoint overhead (for the tropical night sky sucks the soul away from its human moorings and dissolves it in oceans of eternity) he would retell himself the stories he could remember. He believed it was Tycho Brahe, the Danish astronomer and Kepler's teacher, who had introduced Coma Berenices (which in fact was only faintly visible from Cholynesia's latitudes). Instead of choosing a story from Scandinavian mythology, or else making some contemporary allusion, Brahe had instinctively gone back to the classics. The devoted Queen Berenice vowed that if her husband Ptolemy returned safely from war with Assyria she would cut off all her hair and present it to Venus. When Ptolemy duly returned in triumph, she remained true to her promise. Jupiter, awed by this sign of human devotion, retrieved the pledge from the temple and hung her shining tresses in tribute among the spring stars. By such means, Forbici did his best to cling to his culture, his identity, his memories. But that had been in the earlier part of his marooning.

Nights on the atoll were severing one by one the threads which still attached him to a previous life somewhere beyond the flickering horizon. To rest one ear on his gravel pillow was to hear the ocean on all sides of the atoll, rinsing and clucking and mewing. But it was also to hear it thrumming below in deep gasps like those of a labouring beast as it turns and turns a creaking water mill. These sounds filled Forbici with terror. Trapped between the starry ocean overhead and that which surrounded and undermined him, he knew himself about to be engulfed. He awoke from hectic dreams of typhoons, of tsunami-like waves big enough to sweep the atoll bare or else tear it loose like a twirling stone raft. The whole weight of the universe pressed down on him until he felt himself trodden beneath the waves, buried by immensity, sinking through ever-blackening layers, a miserable fragment of mind lost among the ocean's roots. In still other dreams he found himself trudging across waterless orange deserts, maybe on Mars or in some Arabian wilderness, in driven search for the oasis where all yearnings would be quenched. Mirages floated up in the trembling air and flowed together until they formed a rim of palms that encircled his whole horizon. It no longer mattered in which direction he urged his stumbling feet: there was nothing but the mockery of endless recession. Here he would awake in black tears. Lying exhausted, beached on dawn's calm and lapped by undrinkable ocean, he forced his parched mouth to repeat the tales he knew, declensions of Greek verbs, multiplication tables, the names of relatives.

It seems that one day Forbici, delirious with ink, sun and hunger, was sitting semi-comatose beneath his macrophyllum when two heavy wooden craft appeared off the atoll. Long in the water like war canoes, with outriggers and thatched cabins, they were evidently dugouts hewn from monster tree trunks. Besides blood-coloured sails, eight oarsmen apiece propelled them whose polished brown backs glistened with sweat and spray. Each craft had at its prow a striking device: one an enormous fish with gaping mouth and counterbalanced, swivelling eyes, the other a great wicker bird with streaming raffia plumes dyed red, green and yellow.

The two boats hove to within fifty yards of shore. Forbici made no move, having taken them for one more hallucination. He thought this diagnosis confirmed when the fish and the bird began uttering shrill cries, the one rolling its eyes and the other flapping its gaudy wings. He wondered if they were talking amongst themselves or interrogating the islet to see whether it was propitious to land. After a long while the creatures fell silent. Then a hinged flap fell open beneath each, and a child crawled out backwards and fell into the boat; whereupon a drum took up a deep, lugubrious pounding from the shrouded cabin.

These heavy drumbeats began with both drums in unison, slow and regular, appearing to Forbici as the thudding of his own blood in his ears. Little by little the two drummers drifted out of synchrony, becoming increasingly independent and setting up everchanging syncopations. This went on for some considerable time. At length the drummers must have reversed the process, for their rhythms gradually approached one another until the two were finally in unison once more. In his account Forbici observed that it represented a great feat of rhythmic control and musicianship as would far exceed the abilities of even the best European orchestral players. At the time, though, nothing seemed to him more miraculous than anything else because the boundary between a world dominated by ink and one of pure delirium was vague indeed.

As he watched the two strange boats run their keels up the dazzling beach, the state of his head could be described as fugal with assorted wisps of story. The men jumping into the shallows not forty yards away could as easily have been Ulysses and his crew as archipelagic gypsies on the other side of the world. Not knowing whether he was watching events taking place inside or outside his skull, he sat where he was and made no move. At any rate the disembarking men gave no sign of having noticed the presence of a half-cracked European on the islet. They were too much taken up with handing out, from one of the curtained cabins, a life-sized statue carved in ebony. This was carried reverently and shoulder-high beyond the surf in a procession which was brought up by the two little boys (who swam the first few yards), two men carrying the fish and bird figureheads, and, last, the drummers with their turtle-shell instruments.

When the two parties were assembled on the beach, Forbici could see that everyone with the exception of the children was masked. Plain hoods of dark material were drawn over their heads, featureless but for eyeholes.

This procession now walked purposefully across the islet, passing within twenty metres of the observer as he lay beneath his bush. Nobody spoke. The drummers drummed softly, the little boys held hands. The ebony statue was borne above the party's heads at the full stretch of glistening brown arms. They disappeared among the rocks in the direction of the coral pool. It was, as Forbici himself remarked, a measure of his weak and incurious state that he had not the slightest inclination to follow them. In a little while the sound of the distant drumbeats ceased, and all that came to his ears was the sea's familiar stirring. Whether or not all this was a delirious fantasy, it had brought with it a peculiar melancholic charge. The sudden interruption of his long isolation—which he could easily believe had already lasted several years—had imposed a trivial human dimension on what was becoming a grandly geological scene. So far had Forbici turned into a mere component, a kind of sentient rock, that it never occurred to him to announce his presence, to caper with pathetic joy at the sight of fellow humans, or otherwise to greet his rescuers. The waves now spoke to him more familiarly than any human voice. Even when he came to write this day, he was unable to say whether the sadness that had descended so abruptly had to do with the solemnities of the ritual or his sense of being dragged away from an impending private destiny.

Those gull-like cries—were they, too, a property of the moving water that surrounded him? Such mournful screams blurred with surf! He wondered whether he had ever seen a seabird land on this atoll. He could not remember for certain. Long ago he had dreamed of turkey-like birds so tame he could catch them with his hands before roasting and devouring their succulent steaks . . . mere dreams of hunger based on Galapagos tales, that trusting tameness which ought to pose any human being a moral dilemma but which never seemed to. Once broken, the rules of Eden fell in tatters before the rules of self-interest or commerce. Had he not lately read of British entrepreneurs in Antarctic lands? Protestant flint-hearts so weary from clubbing penguins to death that they had taken to driving their sacrificial victims up walkways so that they fell living into the huge cauldrons in which they were rendered down for oil? Again, the far-off cries reached him and, afterwards, a great outburst of drumming.

It may be that Forbici fell into a doze at this point. He next became aware of the scuff of footsteps passing behind his bush towards the beached craft. The masked figures maintained their silence, the turtle-shell drums were mute. The ebony statue was missing, likewise the figureheads and the children. Once embarked, they pushed off. The boats were rowed a few hundred metres offshore before pausing, the one alongside the other as though in discussion. It suddenly mattered to him to know whether or not he was dreaming. He dragged himself from beneath his bush and tottered down to the coral pool on the islet's windward side. There, invisible from the boats, he knelt and peered into the water.

When his eyes had adjusted he could see very clearly the ebony statue lying on the bottom of the shaft among the white pipes of coral. Nearby lay the two figureheads: the great wicker bird with its plumes trailing in the gentle current, the wicker fish with one eye staring fixedly up through the fathoms of pellucid water, mouth agape as though for air. The immense fish he had once thought to trap were bumping at the wicker and ebony with inquisitive snouts.

Forbici crawled back to his shade and discovered that the boats had also returned. The natives had surely not failed to notice his presence even if they had elected not to interrupt their ritual. He never did learn whether the now-unmasked men had come back to rescue him or to salvage the glass carboys, which would have represented valuable storage containers. As they approached the macrophyllum they certainly fell on the empty flasks with excited shouts before greeting this blackened scarecrow of a human with a certain matter-of-factness. Their language was not wholly opaque to him, for the archipelago's dialects overlapped in such a way as to leave audible beneath the surface a framework common to all. He heard '*Aa*,' which he knew to be 'man.' He heard '*bangsa saddi*,' which he interpreted as 'other race.' They half-carried him to one of the boats and gave him musky water from a gourd. It tasted like nectar.

The 'Parergo' describes at some length the weeks he spent aboard a succession of boats as his wandering rescuers went about their maritime business and allowed him to regain his health. They often saw land but seldom went ashore. They did so for fresh water and, on two occasions, for wild pig. But they were clearly not at home there, venturing only fearfully into the jungle which began almost at the high-tide mark. The endless sea was where they felt most secure, and they knew its ways with respectful intimacy. Their fishing abilities were extraordinary, as was their navigation. Out of nowhere, towards dusk, identical boats would appear and moor alongside each other. Sometimes as many as twenty craft congregated in mid-ocean, a family to each, to celebrate a birthday or a wedding. For a while they formed a floating hamlet or random village. Then at dawn, or after a few days, sails would be hoisted and the houses scatter away again over the horizon.

As time passed and language became less of a problem, Forbici was less than ever inclined to return to what and where he had been. This congenial way of living, bounded and dictated by nothing but water, was a transcendent version of the sterile and terminal existence his shipwreck had recently thrust upon him. But the time came when a port was nervously approached and they stood out in the roads until he could be transhipped into a schooner. This was bound for Makassar and the unsought curiosity and attention of his own kind. What comes off these last pages of 'Parergo' is regret. Yet as if Forbici himself could never make his mind up, it is unclear whether this was regret at leaving his rescuers, for their having rescued him in the first place, or even for something yet larger and vaguer he had perceived but could not write. At some point in his mid-ocean stranding on the atoll, he had ceased to be a European, but neither had he become a true sea-gypsy. Maybe that was the source of regret? Maybe, too, he knew he must either lose the sea for ever or else become it for ever by diving in and omitting to surface again. Hence Arezzo, and the grumpy discourse of his later years.

Meanwhile, he learned that the ebony statue had been the body of one of their princes, a kind of sea *datu*, wrapped mummy-like in a single length of sharkskin flayed on the bias in the manner of a peeled orange. Once tightly bound, this skin was given many thick coats of a shellac made from boiled carrageen or similar seaweed mixed with a mucilage extracted from turtle bones and tree sap. This treatment produced the smooth, dully shining black surface he had so easily mistaken for carved wood. The care and ritual the natives had lavished on their chieftain's body was proper to the awe in which they held the sea and their duty to return their own bodies to its dissolving embrace. To lie buried on land was unthinkable. As he understood more and more, Forbici appreciated the irony of having been cast up in a marine cemetery. Not snapped-off flutes of coral, then, but countless gypsy bones had strewn the bottom of the pool. From time to time he heard again the small, gull-like cries among the sound of the waves, and saw the fat fish sniffing around the wicker figureheads lying companionably near the mummy. Had the children known their fate even as they shrilly begged the island's spirits for permission to land, rolling their eyes and flapping their wings? He never found out. Or if he did, he never confided it to his 'Parergo,' with its mood of grieving for unnamed things.

VI

The last pair of divers brought their baskets of lobsters to the surface at around three-thirty in the morning. Both the gullies south and north of the Madonna had been worked out, but there were many crevasses yet to be explored. A couple of empty ice chests were left over when the others had been filled.

'That's enough,' the captain said.

'There's another good hour's darkness left,' somebody objected.

An electrical storm far out to sea gave silent glimpses of cloud banks low on one horizon, snapshots of a bruised pearl-and-ginger colour. Otherwise, the night was lit only by constellations. Little Turtle's river of tears, known elsewhere as the Milky Way, flowed across the sky as though grief were the universe's most prominent feature. It outshone the other playlets, yet the starlight faded as it fell to earth. The ocean mopped up its dazzle as with thick cloth. In the darkness surrounding the *Medevina,* little ambient detail was visible. Whenever stray voltages flickered on the landward quarter, the cape off to port stepped forward and back again indecisively. The coastguard's patrol boats were evidently in another patch of territorial water tonight.

'We shouldn't push our luck.'

The boat's name was a composite of those of Captain Nicomedes and his wife Divina, though the craft was jointly owned by four people who worked it as a cooperative. Medes' decision to call a halt might have been challenged by any of his three companions who thought it wasteful not to profit by an extra hour and fill the last boxes. Yet nobody objected. It was as though everyone aboard was mindful of the dead fisherman we had left afloat so many hours ago, of the vengeance his spirit might wreak if given the provocation. (You will not have forgotten that when the trip was nearly over and a familiar coastline once more off the bow, one of the divers confessed he wouldn't have felt easy that night working any new beds further away from the Madonna, whose calcareous lump had given its blessing to our poaching activities.) Spell and counter-spell.

So the air hoses were coiled and stowed; the engine was stopped and reconnected to the propeller shaft; weak, sweet coffee was passed around in the ringing silence. Cigarette ends glowed in relief. An offshore breeze brought distant beach smells of wrack and drying caves. It was laden also with tropic moulds and that perfume of decay which mixes vegetable with flesh, so the image came of an inflated piglet being rolled at the tideline, trotters stiff among weed. The tension that had preceded the diving had gone, the omens were forgotten. Quiet laughter rose from murmured talk. We were not going to be arrested this trip.

The engine was started and we took up a homeward course. The starlight fell yet more faintly. An overcast was creeping from the east, extinguishing the Octopus's suckers one by one, then the Brazier, and finally blotting Little Turtle's tears. The night grew deeper. Though there was scarcely any wind, the sea became suddenly choppier and from this we knew we had left the shelter of the invisible headland and would soon be out of enemy waters. An hour later, we had a tiny running light aloft, dawn seemed no nearer and the chop was sending smacks of spray from the bows to rattle aft against a hunching of plastic raincoats. This was no threatening change in the weather, though; just the patchy moodiness of the sea as it was strained through the archipelago, bounced between land masses, alternately heated and cooled, convected and churned between abyssal currents and shallow races. Soon the coffee's warmth was long gone and everyone had retreated into himself, wrapped in wet plastic, staring at the same patch of planking or tackle, waiting for the return of daylight and calm water and the implied heat of a rising sun.

The cold end of a working night, the hypnotic blare of the engine: it must have been these which explained the lack of attention. With a crash the boat suddenly reared and flung everyone sprawling in the bilges. Cries went up and the captain cut the engine. The first thought was of having run aground at full throttle, the constant fear in coral seas where hidden reefs can rise to within inches of the surface. But the bottom was still in the boat and we were not yet floundering in the dark ocean. The captain and the mate hurried forward with torches. According to their curses we had rammed a drifting log, a common hazard in these parts.

'If ever a trip was jinxed . . .'

Lights were flashed astern to see what we had overrun. Sure enough, something was rolling in our dispersing wake. The captain was on his knees in the bows, examining the damage. We were holed on the waterline: not a crippling wound, but several square inches of marine ply were stove in. A strong smell of petrol was stealing over the boat. More shouts. A cigarette end was flung judiciously overboard. A spare polythene container of fuel stowed in the bows had been holed by the impact and several gallons had run aft into the bilges before the two men extricated the crushed container and tilted it so that the hole was uppermost.

You will not have forgotten how everyone set to bailing with every available mug and bowl. The boat's rocking dead in the water was accompanied by crunchings and creakings from the darkness where the port outrigger lay. Light-beams focused

384 · JAMES HAMILTON-PATERSON

on it. Tangled up in our paired bamboo floats were the tilted spars of a small boat.

'We hit a boat!'

'It wasn't a log!'

'Jesus, Mary and Joseph! This whole damned ocean's full of abandoned boats tonight.'

'It was a log! Look, you can see it behind us.'

But beneath the vaporous smell of petrol another, familiar stench was filling the air. A single voice cried out, 'Oh my God, it's *him*!,' and a select chill went through us all. The more trembling beams converged on the torpid object in the sea astern, the more clearly they revealed it as the body we had abandoned some eight hours earlier.

If, as claimed, there is in the human mind a gambling instinct, a constant assessment of odds, moments of shock reveal it as being definitely more reliant on superstition than on statistics. There was nobody aboard, yourself included, who was not tugged into the rhetorical device of demanding what were the chances of hitting the same boat . . .? and coming up with the predetermined answer: 'Astronomical!' Afterwards, though, on dry land and with a couple of nights' sleep separating the event from its memory, anyone might soberly have reassessed those chances. The fisherman had been drifting with the current; our captain always went with the current as far as possible. It was the act of a seaman who knows how to lean on nature and save fuel whenever he can. Our collision was unlikely enough, surely, but not so incredible as to make it seem—as it did in the dark with water and petrol bubbling around our bare feet—fated, ordained, even Justice catching up with us. Nobody who lives off the sea remains a rationalist for long. Those of the crew who earlier had consoled their consciences with defiant cries of 'No time!' now began planning the truce they would enact with the dead man before his unappeased spirit might ruin us entirely.

The boat was paddled backwards until his body drifted under the counter. The curious crowded back to view it, thereby making the captain's repairs easier by lifting the hole in the bows clear of the water. The stink of fuel and decay rolled over us. Nobody dared light a cigarette to make breathing easier. Torn-up rags were tied over faces so we became the masked crew of a plague barge. A rope was belayed around one of the fisherman's swollen calves. The noose's friction soon began to slough the skin off, and it seemed likely that his little boat had itself been leaking so that his lower body had steeped in seawater for several hours.

The hatless corpse had swollen greatly since the moment we had last glimpsed him like a lone King Arthur faring forth to Avalon. His back was now as wide and flat as a slab. The T-shirt stretched across it, advertising a brand of paint, was already going in ovals at the seams. His shorts were likewise cracking as though the buttocks beneath were plumped with silicone. Pathos no longer attached itself to this body, wallowing inert as a bolster. His statuesque silence and exaltation had vanished. Only his hair, the liveliest part, floated and streamed in puffs of current as it once had on windy days.

'This time he must be buried.'

'For certain. I don't care if he committed suicide. We're risking our lives until he's given a Christian burial.'

'Oh don't start that again, Mots. We don't know his religion.'

'It's all one God,' the dogged voice said. 'We'll give him a Christian burial because it's the only one we know.'

'You know it, do you? The proper service like Father Deme knows?'

'Even Father Demetrio doesn't necessarily know it. He *reads* it.'

The talk went on. It revealed nothing but a universal refusal to touch the decaying body, to turn it face up, to have anything further to do with it other than piously to wish it gone.

The two boys crawled out on the struts and freed the remains of the fisherman's boat. Without support the little hulk filled and sank until nothing but the outline of its own bamboo outrigger was visible. Since it was too small for an engine, it had never been registered, and there was no identifying number painted on its bow. Nor was there any way of taking it in tow. The captain ordered it cast adrift and it soon vanished.

All this time, dawn was slowly breaking above the cloud cover, sending down a stealing greyness as through a suspended sheet of ground glass. Our world took on shape. A loud swirl of water became visible from just beyond an outrigger.

'*Oy pating, pating!*,' someone shouted, and the additional fear of sharks found a perch next to the other dread spirits already crowding our boat's superstructure. There was no question now of doing anything with the body except sinking it as hastily as possible. The longer it stayed tied to us, the longer we ourselves would remain bait.

Meanwhile, the captain and the mate had jury-rigged a repair. The patch was not quite watertight but, with luck, would hold at reduced speed, provided we kept our weight in the stern. They now decided what they might sacrifice to send the dead man to the bottom. The spare anchor could go, and so could the rusty acetylene cylinder used as a reserve air tank when a third diver was deployed. A discussion broke out as to how much weight was needed to take the corpse down, inflated with gas as it was. The men decided to err on the side of excess. Nobody wanted to see this man ever again although he was probably destined to bob up in many a dream. Now, respectfully, and with the sacrifice of many dollars' worth of equipment, he was being asked to leave us alone.

The assorted weights were firmly tied to the end of the rope attached to his leg. At a suggestion, some people took amulets and crosses from their own necks and wound them around the iron. The captain said: 'I don't know this service. Let's say the *Ama Namin* for the poor fellow, and God give him and us rest.' Out on the dawning ocean the Lord's Prayer rose raggedly towards the ground-glass ceiling. The sea was calmer now, and the boat merely wallowed slightly in the swell. Had the sun been up we would no doubt have been revealed once more as surrounded by an iridescent membrane of fuel clawed here and there into holes by catspaws of wind. Swaying on a few planks above unknown fathoms of water, the crew muttered while the grey bowl of air around us began to fill with the sensation of lifting, as though the remaining darkness were rising like helium to dissolve in the sky. Barely waiting for the various murmurs of 'Amen' and '*Siya nawa*,' the captain began unhitching the cord that secured his precious boat to its albatross. With a series of loud splashes the weights followed each other into the sea.

The length of line gave the iron a second or two in which to sink and take up the slack between it and the dead man's ankle. Then—and how quickly it happened! – the corpse gave a start, a half twist and a sweep of one arm like that of an engrossed snorkeller jerking up for air. For a shutter's click his face turned towards us, a swollen and eroded Buddha, before he vanished. A collective sigh was heaved; nobody spoke. We continued, mesmerised, to watch the water as though expecting him to reappear.

In *The Last Act*, William Tegg quotes a nineteenth-century ship's surgeon, Dr. Clarke, who described leaning out of his cabin window one day idly watching an officer fishing when 'the corpse of a man, newly sewed in a hammock, started half out of the water, and continued its course, with the current, towards the shore. Nothing could be more horrible: its head and shoulders were visible, turning first to one side and then to the other, with a solemn and awful movement, as if impressed with some dreadful secret of the deep, which, from its watery grave, it came upwards to reveal.' In explanation, Dr. Clarke went on to cite one of his medical contemporaries who wrote that 'in a certain stage of putrefaction, the bodies of persons which have been immersed in water, rise to the surface, and in deep water are supported in an erect posture, to the terror of uninstructed spectators. Menacing looks and gestures, and even words, are supplied by the affrighted imagination, with infinite facility, and referred to the horrible apparition.'

So we stood in the lightening dawn and stared at the sea. Hardly any of us doubted the unknown fisherman's power to haunt. If meeting him once had seemed fated, his finding us again had been uncanny. Nobody would have been much surprised, though petrified, had he miraculously started up out of the waves, pointed a dripping finger and anathematised us all. But a long minute passed and nothing more menacing occurred than a couple of abrupt disturbances in the water some way off to suggest the restless prowling of predators. Then the entire boat was enveloped in a reek of corruption that made everyone clap his hands across his cloth-wound mouth. It was so intense it suggested a physical presence, as though the deceased had secretly joined us again, his identity concealed by a mask. On an inspiration a bottle of rum was passed around, the liquor liberally splashed on the cloth protecting mouth and nose. Pulling himself from this collective trance of disgust, the captain tried to start the engine. On the third attempt it caught; and in the welcome blare of machinery we moved slowly from under the primitive cloud and set a course for home.

As soon as the masks were ripped off and lungs were filling with morning air, a discussion broke out between those who believed the appalling stench was the manifestation of an evil spirit and others who thought that when the corpse had reached a sufficient depth, pressure would have ruptured its decomposing tissues and released the pent-up gases in one stinking bubble. Either that, or a shark had attacked it, which would have produced the same result. This debate was still going on when there were fresh cries from the stern and once again the engine was cut.

It seemed that even now the dead man was not finished with us. Someone had noticed that the pile of polystyrene cool boxes containing the lobsters had become shrunken and lopsided beneath its protective shroud of tarpaulin. On investigation it was found that the leaked petrol had melted the bottoms of the two lowest boxes and that all the lobsters in them were dead or contaminated by fumes. One by one the mate began throwing them over the side. Each small splash represented wasted time, wasted danger, wasted effort and diminishing pay for everyone aboard. Finally, the last of the spilt fuel was mopped up and the surviving boxes restacked on a base of boards scrounged from elsewhere in the boat. The two bottomless and soggy boxes were balefully thrown overboard where, light as they were, they drifted cantedly astern as we resumed our course. They remained visible a long while, brilliantly white in the rising sun like children's coffins floating in our wake.

VII

D o you remember this?
Uneasy weather. Beneath blusters of warm wind, the empty metallic ringing of crickets. Uneasy place, too. A little coastline familiar in its every indentation, recently condemned to death, or at least to transfiguration. (The toad developers. Restlessly they gobble up beaches and excrete hotels, restlessly they move on.) You were in a doze, half watching the never-settling terns pick morsels out of the surf when you thought the dead fisherman walked out of the sea and stood in front of the sun. He was bones but he wore the same recognisable clothes and had even found his conical straw hat. The T-shirt was faded, the Boysen paint ad illegible. It was huge now, draped across his clavicles as though on a hanger. The baggy shorts hung off his hip bones. Yet when he began his harangue he enunciated clearly even though his tongue was missing. He was neither maledictory nor exalted, but sneering.

'Easy—it's so *easy*, isn't it? The mellifluous prose? An ear for cadence and some good, heartwarming ethnic action—us brown fisher-folk doing our thing. Add to that the old chestnut, Death, guaranteed to bestow a borrowed solemnity on the emptiest writing.

'So here we are in the South Seas, near enough, which despite everything still carries a feeble voltage of the exotic for your sort. They will ask: *Was it true? Did that story really happen about the dead guy sitting in his boat? Did you actually . . .?*

And there's no answer. That's the splendid thing about tales: it's neither true nor untrue, fact nor fiction. What it is, is words on a page ordered more or less skilfully. You might have witnessed these marine dramas scores of times or never once—there's no knowing.

'Still, there are more bothering things to pick up on, such as why you're so intimate with people you don't fully understand. Why, in short, are foreigners more significant to you than your own kind? I have to assume there's something offensive to you about your own townee culture. No—don't interrupt. You betray an attitude many of your fellows still share more or less unconsciously: that hoary old trope of pastoral virtue as the antidote to urban corruption, itself straight out of the classical authors. You describe the lives and deaths of my own people, fishermen and subsistence farmers, as being in some way "realer" than the so-called conspiracy entered into by your metropolitan man. "Realer" meaning, I suppose, more genuine, more in touch with those damned eternal verities because less padded, less insulated. You appear to be attributing to us a superior moral status which can only be due to our noble poverty even though there's not a man, woman or child among us who wouldn't opt for the contaminating comforts of wealth, given the chance.

'How is it you foreigners feel so free to come half across the world to load us with your fantasies and anxieties when God knows we're laden enough as it is? "Idyll" – that's the word you're always using to describe our workplace. That and "paradise". Look at me' (and here the figure raised his arms, bones greenish in the sun). 'Do I look like someone who has recently passed through paradise? Do *you*? Besides, what's wrong with cement hotels and casinos? There's more future in them than in stick villages. We'd all like the chance of a job in one. Waiter, pool attendant, bell-boy, chef, maid . . . Toiling away, maybe, but toiling away in air-conditioning and for a regular wage. Me, I quite fancy being a croupier. Raking in cash has the edge over hauling in fish.

'Where death's concerned it seems that secularity's blanket doesn't quite cover everything, after all. So in the name of stoicism you invent a peculiar dignity in a corpse being weighted down and thrown into the sea. Let me tell you that my family, whose identity none of you ever bothered to discover, were worried sick about me when I never returned from my fishing trip. Now, of course, they accept I must be dead; but they'd still rather I was lying up in the cemetery where they can visit me instead of having no grave to put flowers on.

'Your attitude to sea burial's just an extension of that other business about virtuous nature being superior to the corrupt metropolis and all its works. A moth-eaten romanticism suggests that sinking gracefully into the mystic deep is somehow more appropriate for a person of sensibility—an ex-person, that is—than being raked out of a crematorium furnace into a jar and then scattered in a suburb. Since you allude to Sir Thomas Browne, you might take his point that it's a waste of time worrying about what happens to your body. It's just your bad luck that you're alive too late to have a soul to fuss over.

'And finally, you patronise everyone when you talk about a conspiracy to pretend death doesn't exist. I can't think why you don't see your urban conspirators as stoical and gallant instead of dismissing them loftily. I tell you, I admire people,' the dead fisherman said. 'I don't care whether they work in boats or stock exchanges—most of the people you meet over the age of thirty have a private loss behind them: dead parent, dead friend, mortal reminders. If only you could stop talking and just observe instead, you might notice that nearly everyone is walking wounded. Yet still they manage to hold it together. We all know what's in store, and aerobics and wallbars are no bad way of flipping it the finger.'

You will remember that it was being able to see the terns again, still not settling at the surf's edge but constantly making as though to alight, that woke you to his sudden absence. The fisherman had gone. Nothing now stood between you and the sun, and in its light the cantankerous and ungrateful old fool's arguments became as shadowy as he. Ungrateful? Certainly. Hadn't every last person on that famous fishing trip contributed to a monument for him? It had become a matter of propriety that none but the crew should have anything to do with it. The result was there for all to see: a modest cement plinth roughly the size of a milestone set among the palms, beyond the reach of any but the most freakish tide. On it was an inscription in the dialect saying 'Remember an unknown fisherman lost at sea and found by God,' then the date. It was signed 'The crew of the *Medevina*.' As for our not having bothered to find out who he was, we had told the headman the moment we reached home, who had informed the village policeman, who had notified the coastguard in town, who had spread the word to all the other coastguards for miles up the coast. We might have kept quiet about it. There were enough long arguments on the way home until the captain wisely decided it was a secret bound to leak out sooner or later. But the real reason for telling people was pure superstition. The crew agreed the dead man's spirit would continue to haunt and jinx them unless he was publicly acknowledged and Father Demetrio held a Mass for him. It was scarcely our fault if, after all that, no news had ever come back about the man's identity. As far as anyone could tell, he was unmissed.

When you considered the dead man's tirade in detail, you thought it vulgar that he made his own kind out to be nothing but yearning materialists. He had implied that poverty reduced everything to mere stark need, abrogating all but the coarsest emotions. Were the poor not to be allowed melancholy, then? Nor the anxieties of love, the arresting queerness of being alive? Are we to deny people their subtlety because they are uneducated? Why shouldn't shore dwellers' tales and traditions be more than fables of crude optimism? Might they not, like other folklore, arouse uneasy doubts about striking it rich or moving away to a distant city in search of fortune, even as they adumbrated love's essential impurity and the heartless equivocations of God?

After all, it is something to be able to sit on a shore beside a driftwood fire while overhead the constellations beam down their implacable playlets and underfoot the skeletons lie in strata yards thick. Briefly sandwiched between two voids, it is something. To lay fish across the coals as your children scamper up from the surf into the circle of firelight long enough to ask, 'Aren't they ready yet?' before vanishing again with squeals: that is not nothing. What it is cannot always be thought of; yet for a flashing moment, the duration of a smile, it feels like the present. There is chatter, and the smoke drifts up, and the children's voices come and go as they play in the dark at the ocean's edge. And you, unhappy mister with the green bones and the smell of fathoms about you, you are dead and not part of this moment. Nor will you ever be again. Maybe that explains the rancour.

The dead fisherman was also mistaken about the significance of sea burial and a corpse's fate. Browne wrote as a Christian; his essay was a declaration of faith. Catafalques and tombs, urns and cenotaphs—of course they were all vanities so long as the souls of the righteous were in the hand of God. Besides, at the sounding of the last trump the dead would be raised incorruptible, attire unspecified (but why not in ragged T-shirt with paint ads?). In any case, it meant one might as well economise on the burial and opt for a simple hole in the ground. But Browne was writing nearly three and a half centuries ago, and such doctrinal optimism is no longer open to us. The fisherman was simply wrong. The whole point of sea burial is that it is *not* monument or grandeur or the survival of even so much as a name. It offers the small consolation of an end which is all of a piece with what went before. By quite simple mathematics it is possible to calculate how, at any moment, the chances are overwhelming that there is in one's lungs a molecule of air that was once breathed by Alexander the Great or Buddha or a dodo or even old Sir Thomas Browne himself (simply name a hero). From there it is a small leap to perceive flesh as wholly pervaded by particles belonging to anyone and everybody. The body is at once atoms of star-stuff and the world's deceased, as well as those of the world's living and the yet unborn. The democracy of this physics is comforting because it requires no leap of faith, only the perception of being a cloudlike formation constantly interchanging parts with friend and stranger, plant and beast, land and sea. Justus Forfex was ahead of his time.

When the terns had gone and sunset's massive city was heaped on the horizon in cinnabar and madder, you will remember that the dead fisherman (even then puffing through your lungs and coursing in your blood) had not had it entirely his own way. Before he faded, you had got a word in, sharp with the triumphalism of the living. You told him there might once have been a pathetic resonance in the suggestion that he had committed suicide, but that you had no further need to be respectful towards a distress that would have vanished at the moment of death. You told him his arguments were nothing, vapour, just the constant opening of blind conversational alleyways which were as constantly closed off again, conceptual clouds. You informed him that you would use him entirely as you saw fit, together with his compatriots, the scenery, diverting cosmogonies, smells, sounds and words. The reason was simple. For as long as you were alive it was your world and your ocean.

VIII

You will remember how puzzled you were as a child to discover that an event is more like itself when anticipated or recalled than when it is happening. The sea's deep discovery must also speak of things other than itself to ring on in the mind, the sonar pulses of inquiry bouncing back from hidden apprehensions. In like fashion, and according to taste, death needs stars and ocean for it to be real, to acquire proper echoes. Cut off from its grand inklings and sprawling starfields, it is wispish. Without playlets it has no weight. Without a little solemnity death is a kind of failure, recasting the handful of days as trivial and forecasting only a savage void. Yet as long as Little Turtle can see it and weep, the fatal life is acknowledged, safe in its moment. The ocean wraps the planet like a caul. Flinging grave, the mind's capacious habitat: we are its children. Provided the sea takes us back, the bloat and decay are acceptable even if the deal remains raw.

Slowly you grew accustomed to this idea that the mind must always falsify things before they can become true. Otherwise, a gap opens between facts and the need to make them private. So you once wrote that even when you were seventeen, you yearned to be seventeen, a fancy that seemed perverse until you found it echoed in Basho's haiku:

Even in Kyoto, hearing the cuckoo's cry,
  I long for Kyoto.

All the unresting melancholy that the now announces—through its cuckoos, its breaking waves, its playground voices, the idea of Kyoto—pushes thought aside. Sitting at night alone on the ocean in a small boat ought to pare things right down. It does; but never to that point of inhuman Buddhistic virtue where one embarks on an immaculate passage into nothingness. The details of survival, of the sea itself, beguile; the playlets captivate. Not being fish, we mourn things, the present especially. The now is as irretrievably shut to fishermen as to poets. It is less accessible even than the past, peopled as that is with ancestors and our lost companions. We are filled with homesickness for no identifiable home, for the Kyoto in which we are writing which is not quite the Kyoto we long to be writing in; for the birthday we passed through but never quite achieved; for the sunlit embrace of an end that will heal us from itself. Playing with time, we embellish the bonework of existence to our solitary taste.

One morning in early spring at an open casement in Arezzo we catch an impossible whiff of ocean. A figure in battered clothes steps out of the front door below carrying a single strapped bag. It is Giusto Forbici. He has the air of someone who will not return. Those cries on a lift of wind from beyond the station are the sound of gulls and surf, or maybe of children who foresee their deaths. That is not snow but petals drifting past the window, the cherries returning to gauntness and waiting for leaf.

Index

pampano, 273, 274.
Panama Canal, 33
pandas, 278.
Panmure, Lord, 66.
'Parallax' (pseudonym), 66–68.
Paramount, Mt., 45, 314.
parrotfish, 273, 287.
passenger pigeon, 281.
Pasteur, Louis. 216.
Payne, E. J., 88 n.
Pearl Harbor, 1, 154, 155, 157–159, 162.
Penguin, HMS, 120, 123, 142, 210, 211, 264, 270, 377.
penguins, as fuel, 271.
Perec, Georges, 101.
perfect pitch, 143.
Permian extinctions, 277.
Pet Travel Scheme (PETS), 110 n.
Peyssonnel, J-A., 119, 120.
pheromones, 140, 364, 331 n.
Phipps, William, 171.
phosphorescence, 223, 323, 325.
see bioluminescence
photosynthesis, 124, 141, 324.
phytoplankton17, 24, 111, 323.
Piccard, Auguste, 221, 222, 234.
pilgrimage, 146, 323.
pistol shrimp, 137.
pit viper, 143.
plate tectonics, 50.
Plato, 100.
Plutarch, 189, 191.
Podestà Is., 92.
pollution, of N. Sea, 24, 170, 239.
of Pacific, 268, 269.
of research bases in Antarctica, 271, 272.
of space, 271, 272.
Porcupine, HMS, 89, 146, 223.
Poseidon, 90.
Power, Adm. Sir A. J., 163, 164.
pressure, effects of, 202–204, 206,

209, 213, 214, 218, 219 221–223.
pressure stocks, 246.
prison islands, 102.
'prochronic' theory, 237.
Ptolemy, 89, 90, 373.
Puerto Galera, 84.
Pyrrho, 67.

Raffles, Sir S., 308.
Reagan, Nancy, 258 n.
Reagan, Ronald, 43, 258 n.
Réaumur, R. A., 119, 120.
Reclaim, HMS, 165.
Red Underwing moth, 282.
Rice, A. L., 13, 217 n.
Richard Montgomery, 173.
Robinson Crusoe, 101.
Rockall, 61, 88, 103.
Roosevelt, Franklin D., 168.
Ross, Sir James Clark, 205.
Ross, Sir John, 67, 206.
Rothman, Brian, 91 n.
Rowbotham, S. B., 66.
Ruskin, John, 238.
Russell, Bertrand, 240.
Russell, E. S., 264.
Rutherford, Ernest, 143.
Rymsdyk, Jan van, 38.

Saint-Exupéry, A. de, 173.
Samal, 299, 300, 303, 304, 324.
Sama sellang (King of the fishes), 325.
Sarah Ann Is., 92.
Savage, V. R. , 308 n.
Scapa Flow, 170.
Schlee, Susan, 121, 215 n.
Schliemann, H., 178.
science fiction, 99, 101.
Scorpion, USS, 178.
Scott, Capt. Robert Falcon, 329.
Scott, Sir Walter, 181.
Scripps Institute of Oceanography, 224, 62 n.

About the Author

James Hamilton-Paterson is the author of several novels, including *Gerontius* (winner of the Whitbread Best First Novel Award), *Cooking with Fernet Branca* (Europa Editions 2005), nominated for the Booker Prize, *Amazing Disgrace* (Europa Editions 2006), and *Rancid Pansies* (Europa Editions 2008). He is the author of two books of poetry and is a regular contributor to *Granta*. He lives and works in Austria.

Carmine Abate
Between Two Seas
"Abate populates this magical novel with a cast of captivating, emotionally complex characters."—*Publishers Weekly*
224 pp • $14.95 • ISBN: 978-1-933372-40-2

Stefano Benni
Margherita Dolce Vita
"A modern fable...hilarious social commentary."—*People*
240 pp • $14.95 • ISBN: 978-1-933372-20-4

Timeskipper
"Thanks to Benni we have a renewed appreciation
of the imagination's ability to free us from our increasingly mundane
surroundings."—*The New York Times*
400 pp • $16.95 • ISBN: 978-1-933372-44-0

Massimo Carlotto
The Goodbye Kiss
"A masterpiece of Italian noir."—*Globe and Mail*
160 pp • $14.95 • ISBN: 978-1-933372-05-1

Death's Dark Abyss
"A remarkable study of corruption and redemption
in a world where revenge is best served ice-cold."
—*Kirkus* (starred review)
160 pp • $14.95 • ISBN: 978-1-933372-18-1

www.europaeditions.com

The Fugitive
"The reigning king of Mediterranean noir."
—*The Boston Phoenix*
176 pp • $14.95 • ISBN: 978-1-933372-25-9

Steve Erickson
Zeroville
"A funny, disturbing, daring and demanding novel—Erickson's best."—*The New York Times*
352 pp • $14.95 • ISBN: 978-1-933372-39-6

Elena Ferrante
The Days of Abandonment
"The raging, torrential voice of [this] author is something rare."—*The New York Times*
192 pp • $14.95 • ISBN: 978-1-933372-00-6

Troubling Love
"Ferrante's polished language belies the rawness of her imagery, which conveys perversity, violence, and bodily functions in ripe detail."—*The New Yorker*
144 pp • $14.95 • ISBN: 978-1-933372-16-7

The Lost Daughter
"A resounding success…Delicate yet daring, precise yet evanescent: it hurts like a cut, and cures like balm."
—*La Repubblica*
144 pp • $14.95 • ISBN: 978-1-933372-42-6

Jane Gardam
Old Filth
"Gardam's novel is an anthology of such bittersweet scenes,
rendered by a novelist at the very top of her form."
—*The New York Times*
304 pp • $14.95 • ISBN: 978-1-933372-13-6

The Queen of the Tambourine
"This is a truly superb and moving novel."
—*The Boston Globe*
272 pp • $14.95 • ISBN: 978-1-933372-36-5

The People on Privilege Hill
"Artful, perfectly judged shifts of mood fill *The People on Privilege
Hill* with an abiding sense of joy."—*The Guardian*
208 pp • $15.95 • ISBN: 978-1-933372-56-3

Alicia Giménez-Bartlett
Dog Day
"Delicado and Garzón prove to be one of the more engaging sleuth
teams to debut in a long time."—*The Washington Post*
320 pp • $14.95 • ISBN: 978-1-933372-14-3

Prime Time Suspect
"A gripping police procedural."—*The Washington Post*
320 pp • $14.95 • ISBN: 978-1-933372-31-0

Death Rites
304 pp • $16.95 • ISBN: 978-1-933372-54-9

Katharina Hacker
The Have-Nots
"Hacker's prose, aided by Atkins's pristine translation, soars [as] she admirably explores modern urban life from the unsettled haves to the desperate have-nots."—*Publishers Weekly*
352 pp • $14.95 • ISBN: 978-1-933372-41-9

Patrick Hamilton
Hangover Square
"Hamilton is a sort of urban Thomas Hardy: always a pleasure to read, and as social historian he is unparalleled."
—Nick Hornby
336 pp • $14.95 • ISBN: 978-1-933372-06-8

James Hamilton-Paterson
Cooking with Fernet Branca
"Irresistible!"—*The Washington Post*
288 pp • $14.95 • ISBN: 978-1-933372-01-3

Amazing Disgrace
"It's loads of fun, light and dazzling as a peacock feather."
—*New York Magazine*
352 pp • $14.95 • ISBN: 978-1-933372-19-8

www.europaeditions.com

Alfred Hayes
The Girl on the Via Flaminia
"Immensely readable."—*The New York Times*
160 pp • $14.95 • ISBN: 978-1-933372-24-2

Jean-Claude Izzo
Total Chaos
"Izzo's Marseilles is ravishing. Every street, cafe
and house has its own character."—*Globe and Mail*
256 pp • $14.95 • ISBN: 978-1-933372-04-4

Chourmo
"A bitter, sad and tender salute to a place equally
impossible to love or to leave."—*Kirkus* (starred review)
256 pp • $14.95 • ISBN: 978-1-933372-17-4

Solea
"[Izzo is] a talented writer who draws from the deep,
dark well of noir."—*The Washington Post*
208 pp • $14.95 • ISBN: 978-1-933372-30-3

The Lost Sailors
"Izzo digs deep into what makes men weep."
—*Time Out New York*
272 pp • $14.95 • ISBN: 978-1-933372-35-8

A Sun for the Dying
"Beautiful, like a black sun, tragic and desperate."—*Le Point*
224 pp • $15.00 • ISBN: 978-1-933372-59-4

Gail Jones
Sorry
"In deft and vivid prose...Jones's gift for conjuring place
and mood rarely falters."—*Times Literary Supplement*
240 pp • $15.95 • ISBN: 978-1-933372-55-6

Matthew F. Jones
Boot Tracks
"I haven't read something that made me empathize with
a bad guy this intensely since I read *In Cold Blood*."
—*The Philadelphia Inquirer*
208 pp • $14.95 • ISBN: 978-1-933372-11-2

Ioanna Karystiani
The Jasmine Isle
"A modern Greek tragedy about love foredoomed, family
life as battlefield, [and] the wisdom and wantonness
of the human heart."—*Kirkus*
288 pp • $14.95 • ISBN: 978-1-933372-10-5

Gene Kerrigan
The Midnight Choir
"The lethal precision of his closing punches leave
quite a lasting mark."—*Entertainment Weekly*
368 pp • $14.95 • ISBN: 978-1-933372-26-6

Little Criminals
"A great story...relentless and brilliant."—Roddy Doyle
352 pp • $16.95 • ISBN: 978-1-933372-43-3

Peter Kocan
Fresh Fields
"A stark, harrowing, yet deeply courageous work
of immense power and magnitude."—*Quadrant*
304 pp • $14.95 • ISBN: 978-1-933372-29-7

The Treatment and The Cure
"A little masterpiece, not only in the history of prison
literature, but in that of literature itself."—*The Bulletin*
256 pp • $15.95 • ISBN: 978-1-933372-45-7

Helmut Krausser
Eros
"Helmut Krausser has succeeded in writing a great
German epochal novel."—*Focus*
352 pp • $16.95 • ISBN: 978-1-933372-58-7

Carlo Lucarelli
Carte Blanche
"Lucarelli proves that the dark and sinister
are better evoked when one opts for unadulterated
grit and grime."—*The San Diego Union-Tribune*
128 pp • $14.95 • ISBN: 978-1-933372-15-0

The Damned Season
"One of the more interesting figures
in crime fiction."—*The Philadelphia Inquirer*
128 pp • $14.95 • ISBN: 978-1-933372-27-3

Via delle Oche
"Lucarelli never loses his perspective on human nature
and its frailties."—*The Guardian*
160 pp • $14.95 • ISBN: 978-1-933372-53-2

Edna Mazya
Love Burns
"Combines the suspense of a murder mystery with
the absurdity of a Woody Allen movie."—*Kirkus*
224 pp • $14.95 • ISBN: 978-1-933372-08-2

Sélim Nassib
I Loved You for Your Voice
"Nassib spins a rhapsodic narrative out of the indissoluble
connection between two creative souls inextricably
bound by their art."—*Kirkus*
272 pp • $14.95 • ISBN: 978-1-933372-07-5

The Palestinian Lover
"A delicate, passionate novel in which history and
life are inextricably entwined."—*RAI Books*
192 pp • $14.95 • ISBN: 978-1-933372-23-5

Alessandro Piperno
The Worst Intentions
"A coruscating mixture of satire, family epic, Proustian
meditation, and erotomaniacal farce."—*The New Yorker*
320 pp • $14.95 • ISBN: 978-1-933372-33-4

Benjamin Tammuz
Minotaur
"A novel about the expectations and compromises that humans create for themselves...Very much in the manner of William Faulkner and Lawrence Durrell."—*The New York Times*
192 pp • $14.95 • ISBN: 978-1-933372-02-0

Chad Taylor
Departure Lounge
"There's so much pleasure and bafflement to be derived from this thriller by novelist Chad Taylor."—*The Chicago Tribune*
176 pp • $14.95 • ISBN: 978-1-933372-09-9

Roma Tearne
Mosquito
"A lovely, vividly described novel."—*The Times* (London)
352 pp • $16.95 • ISBN: 978-1-933372-57-0

Christa Wolf
One Day a Year
"This remarkable book offers insight into the mind behind the public figure."— *The New Yorker*
640 pp • $16.95 • ISBN: 978-1-933372-22-8